X-
RAY

Corporeal Matters
is a publication series on arts-based research, in particular practices and concepts that place the body at the core. It illuminates how the body appears simultaneously as witness, document, and agent in contemporary life, and offers insights into corporeality as the often-neglected dimension that cuts through ethics, aesthetics, and politics. From multiple perspectives and fields of application and grounded in moments of research, encounter and debate generated in the context of the Inter-University Centre for Dance Berlin (HZT), the series hosts edited volumes, authored publications, workbooks and other formats.

Series Editors
Janez Janša, Sandra Noeth and Sandra Umathum

THE GAZE OF THE X-RAY

An Archive of Violence

Shahram Khosravi (ed.)
2024

[transcript]

Content

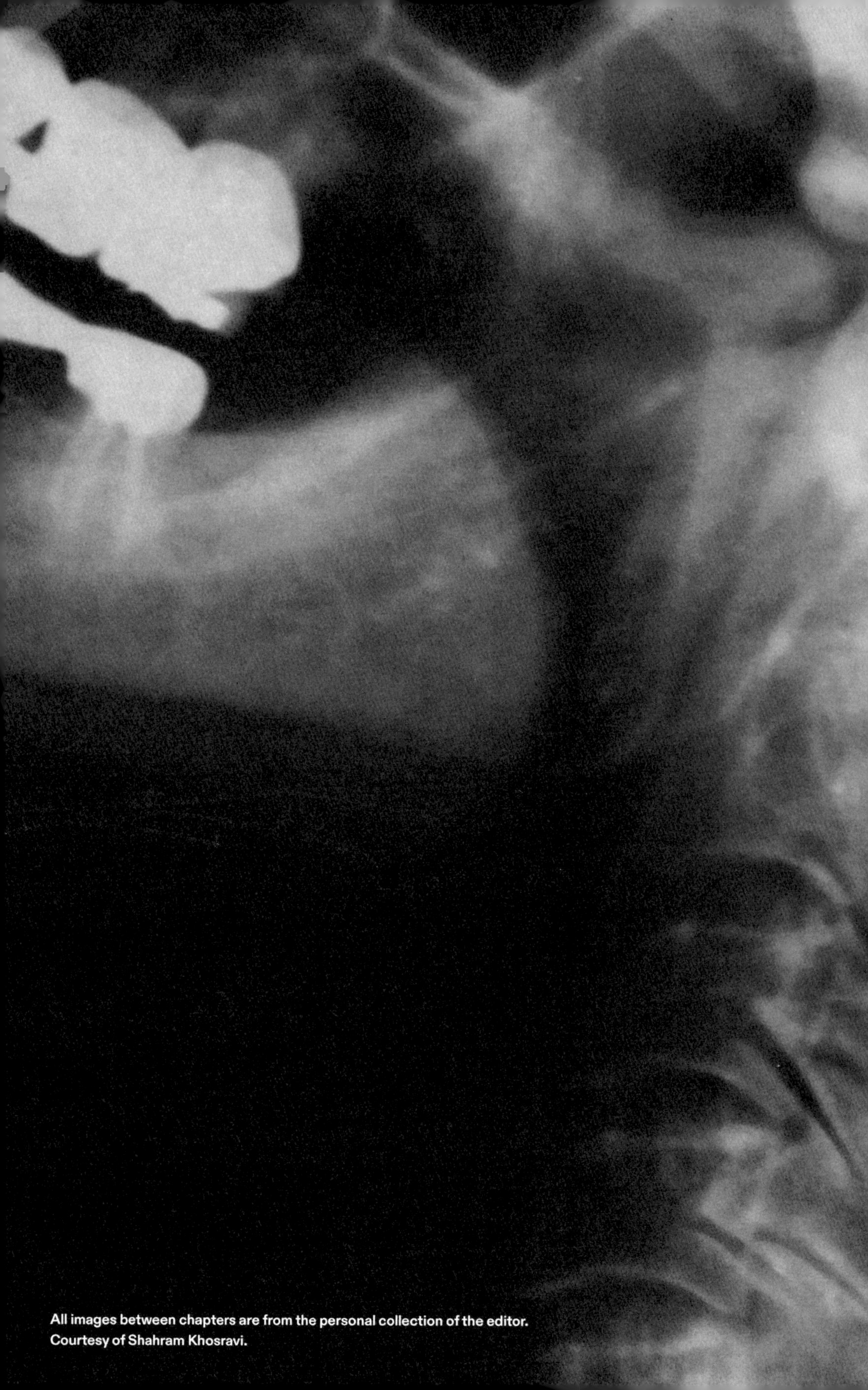

All images between chapters are from the personal collection of the editor. Courtesy of Shahram Khosravi.

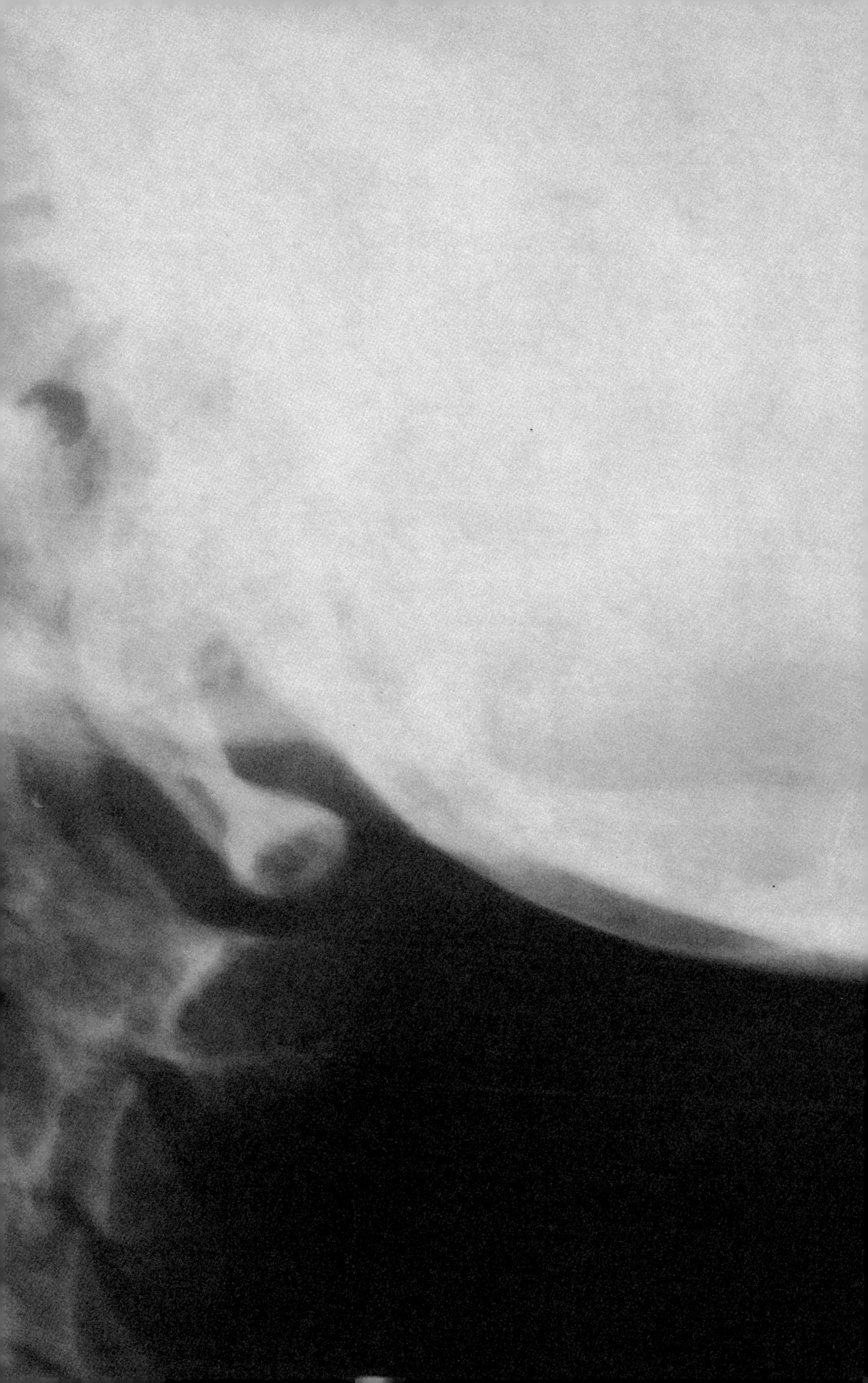

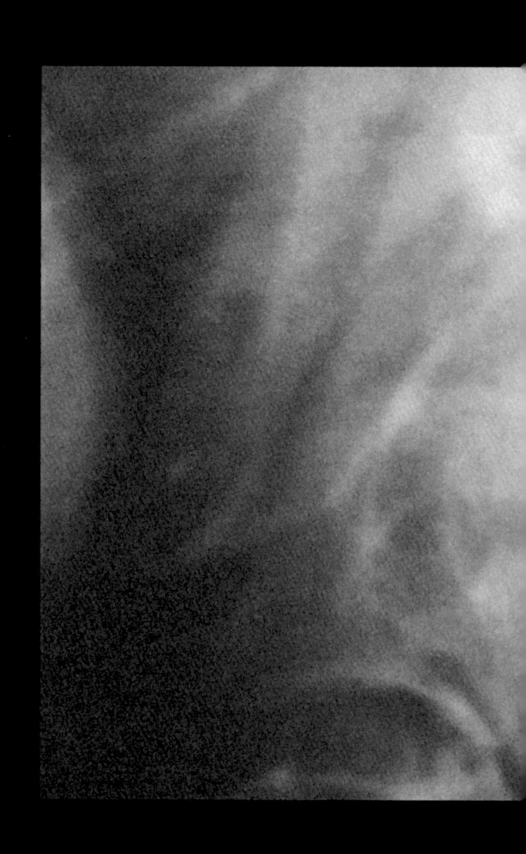

Shahram Khosravi

Wayward X-ray Photos

Turning the body into evidence has a long history. However, the brutal methods of surveillance of bodies reached new levels during slavery. Techniques of measuring, documenting and monitoring captive bodies developed. In the second half of the 19th century, thanks to the invention of new technologies, modern methods of screening bodies expanded. This was when Alphonse Bertillon designed anthropometric measurements for bodily identification of criminals in France. A couple of decades earlier, the British man William Herschel, governor and judge in colonial India, suggested fingerprinting as the optimal method for identifying individuals. Fingerprinting was used as a technique to suppress local rebellion and what the colonial authorities called "criminal tribes."[1]

During the same period, in the second half of the 19th century, through a series of inventions, the ways in which we perceived our senses changed dramatically. These inventions were photography, cinema, the gramophone, and the X-ray. Since its invention in the late 19th century, the penetrating gaze of the X-ray has changed our vision of the inside of the human body. After we started to see inside ourselves, the relationship between ourselves and our bodies changed forever. Almost during this same period, modern urban planning and architecture developed. While the X-ray allowed you to see inside the body,

[1] Chandak Sengoopta, *Imprint of the Raj: How Fingerprinting Was Born in Colonial India* (London: Pan, 2004).

modern architecture offered the viewer the ability to see inside buildings by revealing their interior.[2]

X-ray technology developed during a period of time that was characterized by the demand for total transparency and knowability, partly as a colonial practice and partly as part of the governmental apparatus of the emerging modern nation-states. The 'X' in 'X-ray' refers to the unknown and to the mysterious. The X-ray is supposed to make visible what is invisible and to reveal what is concealed. Used originally for medical diagnosis, it soon became a technique for surveillance and policing, in which the central desire has been to 'see through' human beings.

Accordingly, a new regime of visibility emerged. A penetrating gaze demands transparency and nakedness. The penetrating gaze stretching over time and space—from the Lantern Law in late 18th century New York, which required all slaves to carry a lighted candle when going out at night; to the vertical gaze of American drones over the occupied lands in the Middle East; to surveillance by algorithm; to the European Union's asylum seeker fingerprint database (EURODAC)—all these eyes upon eyes enable those in power to penetrate into the lives, souls, and bodies of the Other. This assemblage of eyes upon eyes creates a monster not unlike the Greek myth of *Argus Panoptes*, a hundred-eyed monster, from whom the term panopticon is derived.[3]

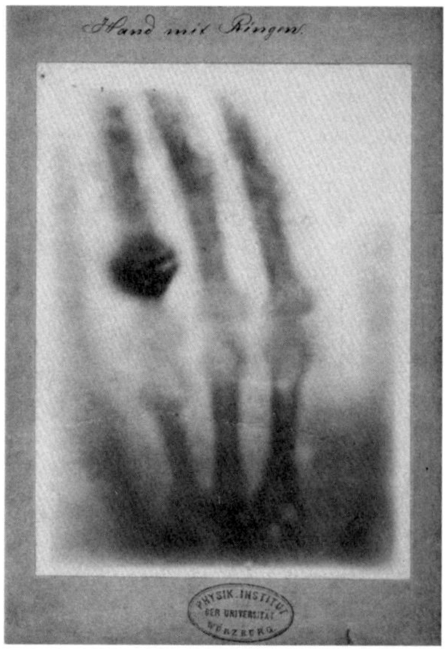

Fig. 1 — First X-ray photo. Anna Bertha Ludwig's hand, 1895.

The new regime of visibility has created a complex apparatus of illuminating the otherness of racialized, sexualized, and gendered bodies.[4] All to render the Other transparent and knowable. The will to know is not in the sense of recognition of the Other, but in the sense of grasping, of getting hold of. The intent is to dominate. In making the Other knowable, her body is smashed and fragmented into measurable data pieces. Making her knowable makes her de-faced, de-named, ghostly, like a figure in an X-ray photo. Common to all these technologies is that the individual is deprived of a human face, while

[2] Beatriz Colomina, *X-Ray Architecture* (Zürich: Lars Müller Publishers, 2019).

[3] Shahram Khosravi, "Bordered Imagination", *Infrastructural Love* (Basel: Birkhäuser Verlag GmbH, 2022), 297–305.

[4] Simone Browne, *Dark Matters: On the Surveillance of Blackness* (Durham, NC: Duke University Press, 2015).

everything about her (body, background, person) is screened, simplified, documented, and explained. The Other's body is both hyper-visible and invisible at the same time.[5]

During its development, X-ray technology fitted well into the regime of visibility and its demand for transparency and nakedness. Offering corporeal transparency, the X-ray has become a crucial part of regulating bodies in modern society. However, the interest for the modern medical vision of the X-ray is rooted in the theatricality of anatomical medicine during the Renaissance. The past's spiritual interest for the body's interior is replaced by a modern biopolitical interest in monitoring the inside as well as the outside of the body. While in the past the interior of the human body was searched for traces of divine transcendence, modern techniques reduce the body to a mere body-object.

In late 1895, Wilhelm Conrad Röntgen realized that he had discovered something extraordinary. He asked his wife Anna Bertha Ludwig to come to his laboratory and the first X-ray photo was produced; in it are the bones of her left hand with a wedding ring on the second finger from left. On seeing the skeleton of her own hand, she screamed, *"I have seen my death!"*

The fascination to see inside the body caused a kind of 'X-ray mania' among the rich. In these early X-ray images, we see the hands of many kings, queens, and aristocrats. Interestingly in the case of women, the wedding rings, signs of family values and wealth, are always present in the photos. X-ray photos soon became "intimate photographs" that lovers gave each other. In Thomas Mann's classic novel, *The Magic Mountain*, the protagonists carry X-ray photos of their lovers in their pockets, and Claudia, one of the main characters, before leaving the sanatorium, gives her lover an X-ray photo of herself.[6]

It is no coincidence that the female body became the main target of this X-ray mania. X-ray is part of a broader visual field that has never been neutral. The gaze is a hierarchical and interwoven complex of gender, race, and class. The X-ray became part of male voyeurism, a desire to look inside the female body, to expose

[5] Frantz Fanon, *Black Skin, White Masks* (New York: Grove Press, 1967), 109.

[6] Thomas Mann, *The Magic Mountain* (United Kingdom: Vintage Books, 1969).

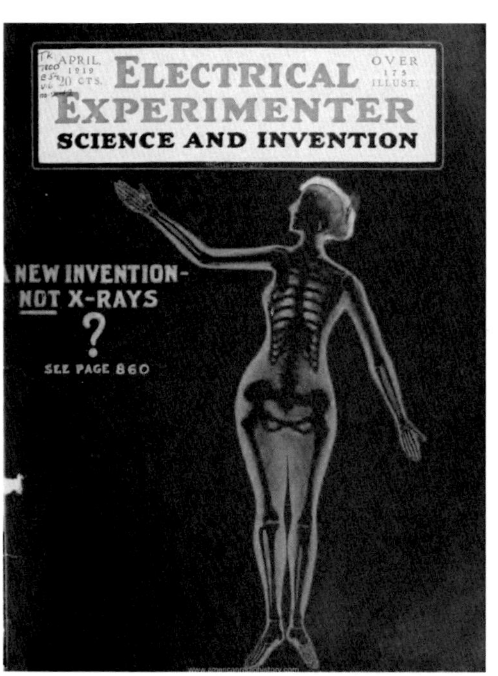

Fig. 2 — Electrical Experimenter, April 1919.

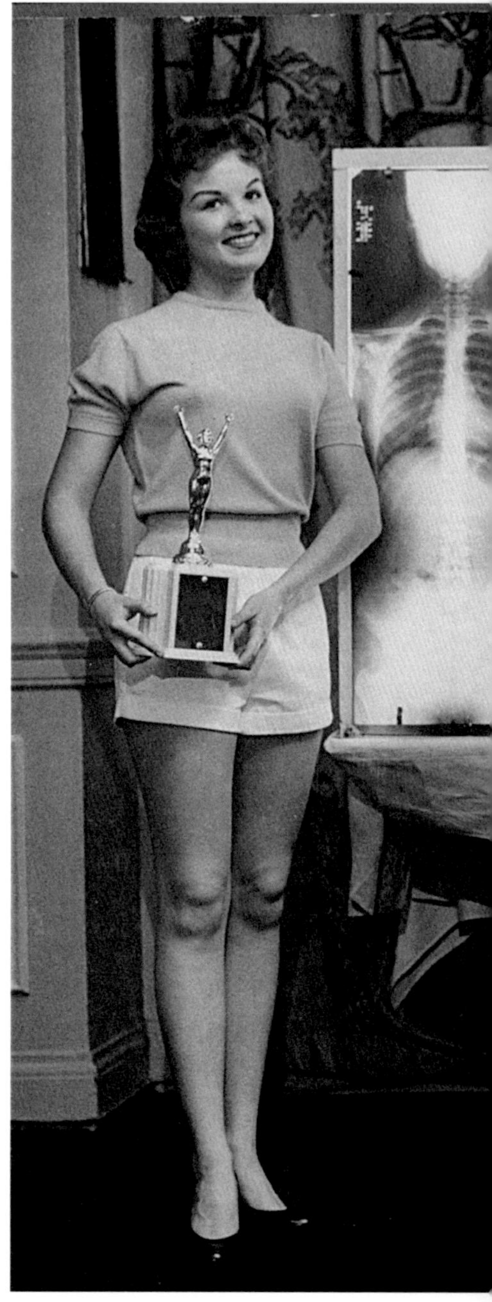

Fig. 3 — Wallace Kirkland/The LIFE Beauty Contest, Chicago, 1956.

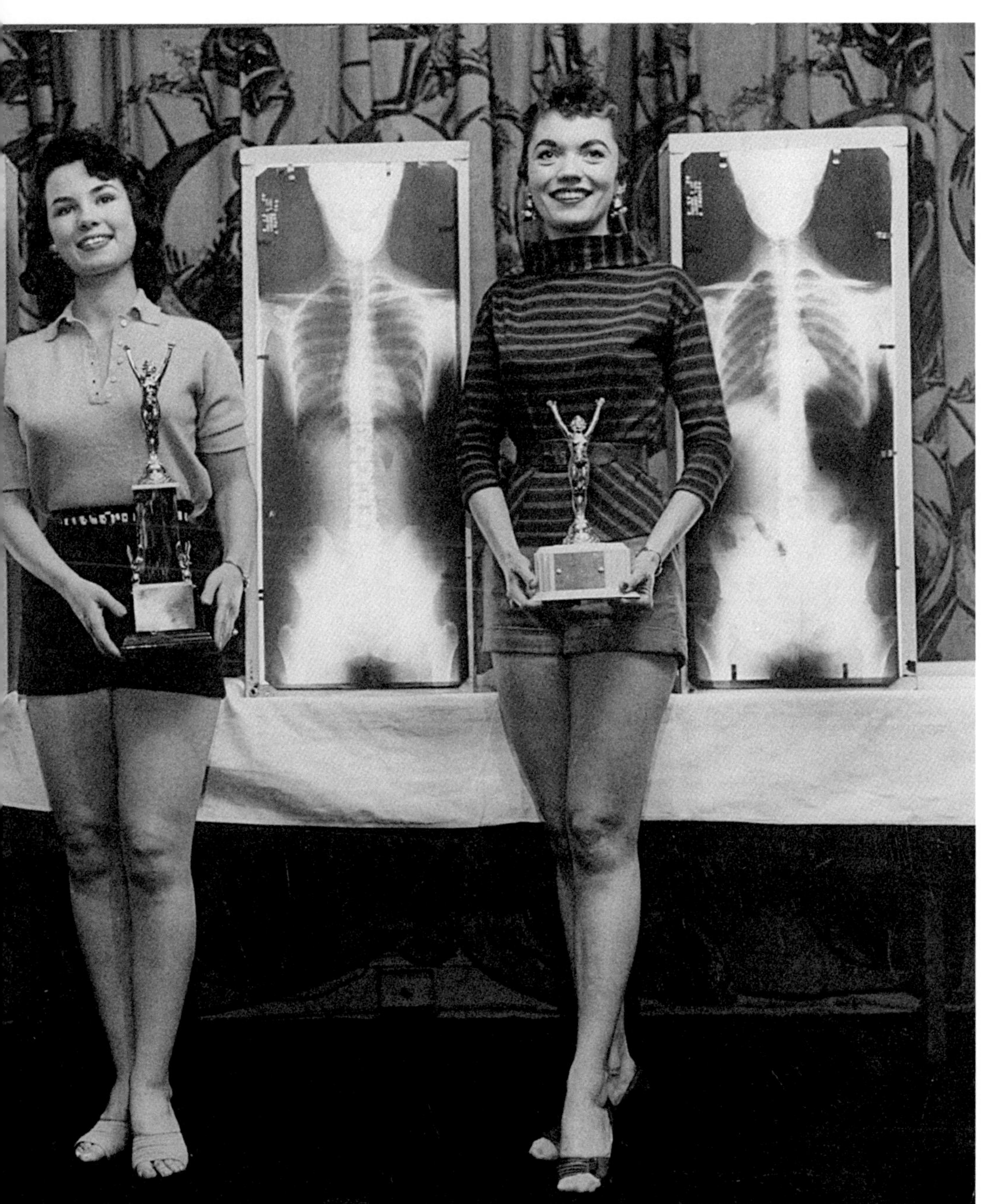

the private parts of it. The modern medical gaze has contributed to men's exclusive access to women's bodies. Through the larger part of the last century, X-ray technology became part of controlling, disciplining and commodifying the female body.[7] Mass radiography campaigns during the 1940s were used for detecting ailments such as tuberculosis. However, women were X-rayed differently than men.[8] Regarded as more prone to infectious diseases such as tuberculosis than men, women were treated as carriers of diseases and therefore their bodies were more extensively screened and monitored. The male erotic lure of X-rays and its perverse gaze constituted an object of desire, inside as well as outside the female body. The full body scanners used at airports are known as "porno scanners" in the security industry.[9] Even the expansion of X-ray mammography after World War II gradually became part of the construction of what is assumed to be 'normal breasts.' An X-ray photo is not an image of the body, it is an image of the body being imagined.[10] The male erotic imagination used the X-ray to create imagined physical fitness and the perfect figure of women. During the 1950s and 1960s, beauty contests in the US used X-rays to judge women's 'inner beauty'; i.e. a straight spine and skeletal symmetry. There is a continuation between the X-ray mania among European aristocrats at the turn of 20th century and the development of beauty standards among the bourgeoisie in the second half of the last century.

However, while X-ray technology was used in order to commodify the female body within bourgeois consumer culture (see Paolo Favero's essay in this volume), during the century, X-ray practices were part of regulating and criminalizing bodies of poor and Black bodies. X-ray technology, as part of the modern regime of visibility not only contributed to the production of the racial, sexual, and poor Other, it also contributed to the

[7] Lisa Cartwright, *Screening the Body: Tracing Medicine's Visual Culture* (Minneapolis, MN: University of Minnesota Press, 1995). See also Evan Smith, and Marinella Marmo, "Uncovering the 'Virginity Testing' Controversy in the National Archives: The Intersectionality of Discrimination in British Immigration History", *Gender & History* 23, no. 1 (2011), 147–65.

[8] Catherine Waldby, "Medical Imaging: The Biopolitics of Visibility", *Health* 2, no. 3 (1 July 1998), 372–84.

[9] Frances Stonor Saunders, "Where on Earth Are You?", *London Review of Books*, 3 March 2016, 7–12. Accessed 30 September 2023.

[10] Cf. Cartwright, *Screening the Body*, 159.

emergence of the able-bodied white bourgeoisie. The juxtaposition of medical imaging and media technologies have led to the mediation of human bodies.[11] Since technologies are designed and produced in social and historical contexts, what they produce—X-ray photos—are part of the broader social stratification. X-ray technology has not only produced a gendered and sexualized gaze but a racialized one as well. As a medical technology, the X-ray was used not only to reinforce racial imaginaries about the Black body but has also been used against it. During the first decades after its invention, X-rays were used for bleaching in the US, by removing skin pigmentation.[12] In line with other inspirations the Nazis took from American racism, the X-ray was vastly used in Auschwitz in medical experiments for sterilization.[13] The racist belief that Black people have denser bones, harder flesh, and thicker skin from the 18th and 19th centuries was still widespread among 'medical scientists' in the second half of the 20th century. Accordingly, Black people were X-rayed differently than white people. Until the 1970s, X-ray technicians used higher radiation doses for Black bodies.[14]

X-rays were a good fit in anti-poor stances and practices. While in the early period of X-ray mania European aristocrats took X-ray photos of their hands with diamond rings, later on in South Africa, X-ray technology was used to detect diamonds hidden in the bodies of mine workers. Every day after they finished work in diamond mines, Black workers queued to be screened by X-ray machines. White experts were hired to monitor the inside of the workers' bodies to find any hidden diamonds that the workers attempted to smuggle out. The workers were presumed guilty until proven innocent. Simone Browne examines the 'surveillance' of blackness from the slave ships to the present.

[11] José van Dijck, *The Transparent Body: A Cultural Analysis of Medical Imaging* (Seattle, WA: University of Washington Press, 2005).

[12] Carolyn Thomas de la Peña, "'Bleaching the Ethiopians': Desegregating Race and Technology through Early X-Ray Experiments", *Technology and Culture* 47, no. 1 (2006), 27–55.

[13] Susan Benedict and Jane M. Georges, "Nurses and the Sterilization Experiments of Auschwitz: A Postmodernist Perspective", *Nursing Inquiry* 13, no. 4 (2006), 277–88.

[14] Itai Bavli, and David S. Jones, "Race Correction and the X-Ray Machine — The Controversy over Increased Radiation Doses for Black Americans in 1968", *New England Journal of Medicine* 387, no. 10 (8 September 2022), 947–52.

862 ELECTRICAL EXPERIMENTER April, 1919

Locating Stolen Diamonds by X-Rays

Possibly you will remember having read from time to time of the remarkable tricks resorted to by the native diamond miners in the great Kimberly diamond region in South Africa and other parts of the world. So great has the temptation often become to steal diamonds, especially when an extra large one may have been suddenly unearthed, that these natives have been known to resort to the most unbelievable tactics in order to carry the diamonds out of the mine and to withstand inspection even when stript, as practically all of them are, before they leave the mine at the end of the day's labor.

One of the successful schemes which has been worked out by the superintendent of a large South African diamond mine is shown in the accompanying illustration, and it involves the use of a powerful X-ray machine having several X-ray bulbs excited simultaneously. As each miner passes before the X-ray bulbs, the examiner looks thru his fluoroscope and rapidly swings it up and down, so as to take in the entire body in a few seconds. This system of detecting the presence of a diamond, no matter whether it is buried in the flesh, resting in a throat cavity, or even in the stomach—an almost unbelievable practise resorted to in several instances on record—the X-ray examination quickly indicates the presence of the diamond.

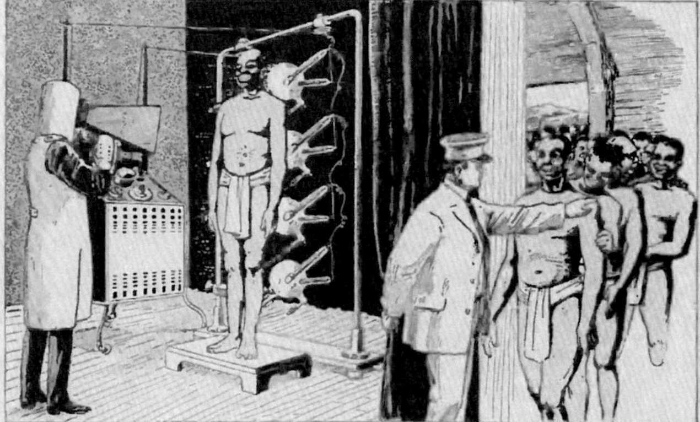

The Tricks Resorted to by the Native Diamond Miners in Kimberly, South Africa, and Other Mines, Pass All Human Belief and Imagination at Times. Cases Have Been Known Where the Lucky Finder of a Particularly Fine Specimen Even Swallowed the Stone, Intending Presumably to Regain the Diamond Later. In Some Instances Diamonds Have Been Secreted in Self-Inflicted Wounds or Incisions in the Leg. But the X-Ray Spoiled All These Clever Ruses as Soon as It Was Adopted for Examining the Miners Every Day, Before They left the Mines. The Eye of the X-Ray Sees All.

Of course, the logical question that arises is—How can the X-ray detect the presence of a diamond inside of the body; especially when it may be temporarily lodged by the clever thief in proximity to large or fairly large bone structures, which would seem to preclude any possibility of detecting the precious stone? However, a perusal of a table showing the various transparencies of different materials under the X-ray will give the solution to the problem. It has been found that the diamond has a different transparency than any ordinary materials, including the bone and flesh of the body, which might happen to be in proximity to it at the time of such an X-ray examination. Also the diamond is a most peculiar substance, and it has certain fluorescent properties which render the facility of its detection all the more possible under an examination by X-ray, as it has a tendency to fluoresce or glow slightly when under the influence of X-rays, which phenomenon is readily detected on a sensitive fluorscope or X-ray screen.

The X-ray machine here shown is connected to a battery of four powerful X-ray tubes of the latest Coolidge type, as otherwise if the tube had to be moved up and down behind the subject, considerable time would be lost in performing this operation, and where several hundred subjects have to be examined in a very short space of time, it can readily be imagined that such a device as here shown is imperative.

GIANT SUBMARINES HAVE 12-INCH GUNS AND STEAM PROPULSION.

We are now able to publish a photograph of one of the most jealously guarded secrets of the British Navy. While the Germans were boasting of the huge under-sea cruisers with which they proposed to gain the control of the seas, the British Admiralty were constructing submarines capable of matching the largest destroyers afloat and of fighting even cruisers in a surface contest.

The secret of these boats was their great size and speed and the fact that while on the surface they used steam as their propelling power, carrying two funnels like an ordinary surface warship. One photo shows a British "K" class, two funnel submarine "steaming" on the surface at sea. This is the largest class submarine produced by any nation and is 340 feet in length. It outclasses any U-boat built by Germany. Great Britain has a whole fleet of these sea terrors. Storage batteries and motors are used while running submerged.

The other photo shows a new British monitor submarine with a 12-inch gun, capable of giving battle to most any class of armed ship under favorable conditions. So far as known this is the first photograph to be received in this country showing Great Britain's combination of the U-boat and coast defense vessel. The 12-inch gun is the largest that was ever mounted on a submarine.

Fig. 4 — Electrical Experimenter, April 1919.

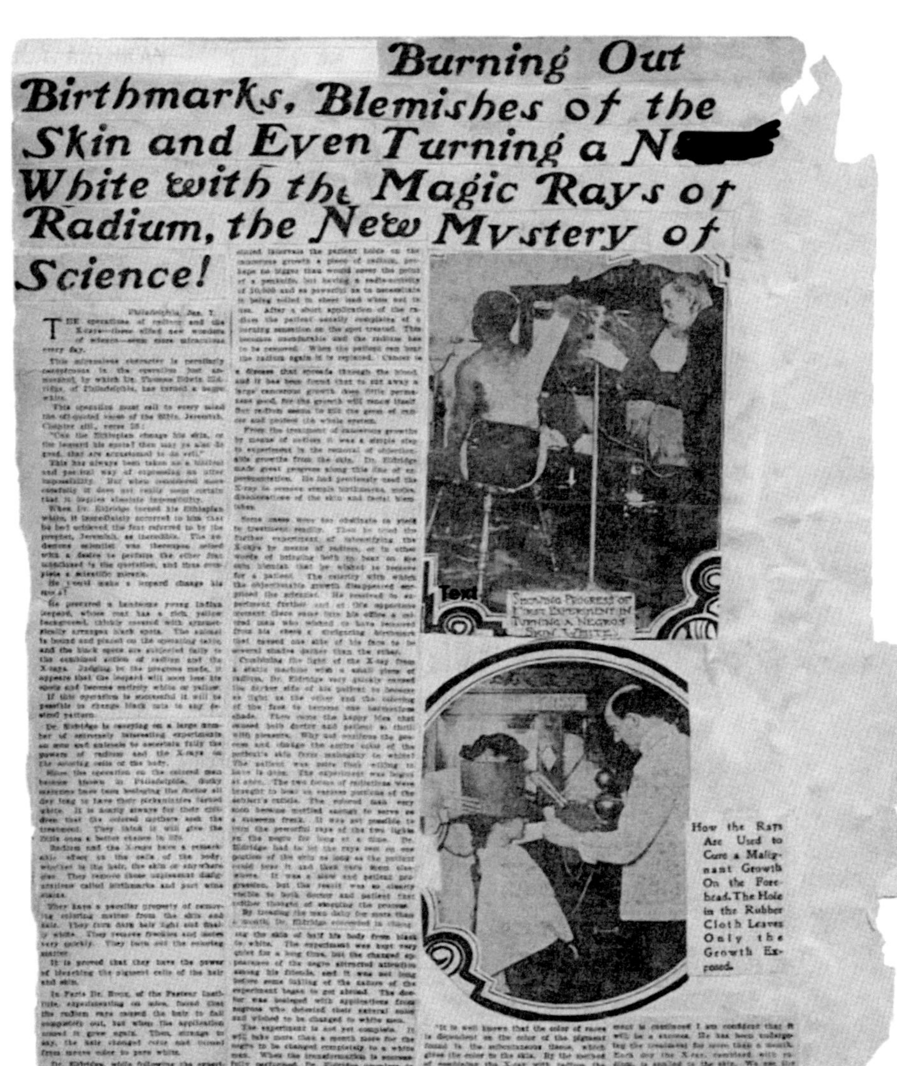

Burning Out Birthmarks, Blemishes of the Skin and Even Turning a N[redacted] White with the Magic Rays of Radium, the New Mystery of Science!

Fig. 5 — Boston Globe, 25 January 1904.

She traces how "an ontological link between labour preparedness, race, ethnicity, and resistance" was established.[15] The regime of visibility has principally been a method of administration of the labour of Black workers.

Borders

This culture of criminalisation of poor people and the non-white body is still widely exercised against asylum-seeking minors. Unaccompanied minors who seek asylum in Europe are exposed to X-ray surveillance for age assessment. They are presumed to have lied until proven innocent. In both cases, both of the diamond mine workers and the age assessment of underage asylum seekers, the X-ray photos are 'convict images'. Like the 'convict images' of prisoners, X-ray photos belong to a visual genre of criminal identification, defining the criminalised body. They are taken without consent. Even if, as in the case of asylum seekers, it is offered as 'an option', in practice it is not, since the consequence of not accepting the imposed age assessment will be deportation. Thus, X-ray technology has been used to provide evidence of 'crimes' and 'untruths' in order to punish Black people, workers, border transgressors, and asylum seekers. Moreover, it has established an uneven regime of transparency.

For poor mine workers or underage asylum seekers, X-rays have been a technique to externalize an 'inner truth'. It is a Foucauldian 'regime of truth', i.e. the truth is constituted through X-rays. X-ray photos speak for themselves. The asylum seeker is defined with the help of X-ray photos of her wrist bones or teeth. There is no need for her body. No need to hear her own words or know why she is seeking asylum. X-rays determine if she is old enough to be deported. Interestingly, the asylum seeker's body is measured and is defined in relation to bodies that are geographically and temporally distant. For instance, in French border practice, the main reference for age assessments is the *Greulich and Pyle Atlas*. The atlas was designed for skeletal age readings based on a collection of X-rays of 1000

[15] Browne, *Dark Matters*, 95.

American white youths in Ohio, taken between 1931 and 1942. X-rays of the wrists of asylum seekers today in France are assessed with reference to this atlas.[16] The power of the X-ray as a technique for extracting 'the truth' stretches across space and time.

The interlinking of modern medical technology with a regime of surveillance reflects the ancient Roman common law *Habeas corpus,* which literally means, "You shall have the body." The politics of life is the politics of the body and the bordering practices dictate having the asylum seeker's body to extract 'the truth' from it. By splitting the body into data pieces, X-ray technology turns the body of the 'suspects' into evidence against them. The moment one realizes that one's own body has been consumed within a power structure against oneself, is the moment of separation between the body and its representation.

The cases mentioned previously show how, with the help of X-rays, women, Black mine workers, and asylum seekers have been reduced to merely body-objects, a thing to be commodified, dispossessed, and punished. Moreover, as we see in the essays by Reza Hussaini and Jonathan Krämer in this book, medical technologies have been used in border control processes in other ways as well. Health screening has been and still is a compulsory part of the visa application process. Applicants from poor countries desiring a visa to enter rich countries are required to undergo a chest X-ray as a health examination and in particular to check for tuberculosis. Migrant workers in the United Arab Emirates are regularly forced through health screenings and failed tests result in immediate deportation.

The technique of oppression uses the body of the marginalised against her. As a medical technology, the X-ray has been a tool for saving lives but at the same time it has been a device of control and oppression. How X-ray technology has been used depicts the symbiosis of violence and care. The same technology used for saving lives has been used to expose certain bodies to the culture of disbelief and criminalisation. While the Renaissance theatricality of anatomy used the bodies of the poor, executed 'criminals'

[16] Sandrine Musso, "The Truth of the Body as Controversial Evidence: An Investigation into Age Assessments of Migrant Minors in France", in *Waiting and the Temporalities of Irregular Migration* (London, New York: Routledge, 2020), 151–169.

and colonized bodies, the medical gaze of the X-ray has been used to monitor the inside of the bodies of asylum seekers (age assessment), migrant workers (visa application), women (sexual commodification), and mine workers (criminalization). By providing see-through vision, the X-ray is expected to guarantee more security. Not surprisingly, the label 'X-ray' has been used to name security operations. The notorious detention facility at Guantanamo Bay was named *Camp X-Ray,* and the racist police operations to arrest and deport undocumented migrants in Thailand was named *X-ray Outlaw Foreigners.* The expansion of the security industry as part of the global war on terror has led to the proliferation of X-ray scanner technology. Airports, ports, and border check points all are equipped with small and large X-ray machines.
It is a fast and efficient way to inspect bodies and cargo when crossing borders. The paradigmatic image of the world today is undoubtedly a picture of bodies squeezed between pallets inside a truck. The picture is taken by a cargo X-ray scanning camera on the border between two nation-states. The X-ray image shows the naked white bodies on a black background—a silhouette of human beings. The human body is shattered into data pieces, defaced, and turned into a ghost.

The Ghosts

As Sandra Noeth and Joanne 'Bob' Whalley write in their essay (this volume) X-raying "turns the body into a promise. A promise of wholeness and integrity." X-ray photos are future oriented. They tell us something about the future, i.e. the decay of the body. Some are documents of ruination. They both document ruination and the violence that ruins. Ruination is an ongoing process that is related to precarity, vulnerability, dispossession and decay. X-ray photos also recall death. The skeleton in the X-ray photo reminds us what will remain after death. Recall Anna Bertha Ludwig's scream upon seeing the first X-ray image featuring the skeleton of her hand, in X-ray photos we see our death.
Looking at X-ray photos is an act of haunting. The ghosts, not only from a haunted past but also from a haunted future, are dialectical ghosts who link the past and the future to the present. As Avery F. Gordon sees it, ghosts are social figures

whose testimonies are necessary in order to hear concealed narratives.[17] The ghosts of X-ray photos bear witness to violence. We need to admit the ghost into our analyses of social and political life. We need to call upon them as witnesses to visualize and verbalize that which has been unseen and unspoken in the official archives, histories, and narratives.

Why an X-ray Archive?

X-ray images capture the past. They register what has already happened, a damage, an injury, but at the same time, they are expected to reveal something about the future; the possible scenarios of a health condition. Moreover, X-ray technology also promises the potential for detecting other yet-to-happen crimes: transgressions and dangers; terrorism, smuggling or irregular migration.

In X-ray photos the past and future are bound together. This is how archives function too. In *Archive Fever*, Jacques Derrida presents the archive rather than a thing of the past, the

> question of the future, the question of the future itself, the question of a response, of a promise and of a responsibility for tomorrow.[18]

The response-ability of the archive renders the archive open to the future. Ariella Azoulay sees the archive not as a collection of isolated fragments from a completed past but rather as a space of shared life where the past is always incomplete and the present always in becoming.[19] This is what Walter Benjamin referred to as the "incompleteness of the past."[20]

[17] Avery F. Gordon, *Ghostly Matters: Haunting and the Sociological Imagination* (Minneapolis, MN: University of Minnesota Press, 1997).

[18] Jacques Derrida, *Archive Fever: A Freudian Impression*, trans. Eric Prenowitz, Religion and Postmodernism (Chicago, IL: University of Chicago Press, 1996), 36.

[19] Ariella Azoulay, "Archive", *Political Concepts, a Critical Lexicon*, 21 July 2017, https://www.politicalconcepts.org/archive-ariella-azoulay/. Accessed 30 September 2023.

[20] Walter Benjamin, "Theses on the Philosophy of History", *Walter Benjamin: Selected Writings, 4: 1938–1940* (Cambridge, MA: Harvard University Press, 2006), 389-400.

The X-ray archive is an attempt to visualise the traces and the pattern of violence, practised by the states, racial capitalism, colonial racism and sexism. By juxtaposing different cases across time and space, this archive demonstrates a set of relations. The relationality in this archive does not emerge from globalisation and transnational trade, media or tourism, but from the pain of oppression. This relationality reveals multiplicities of violence in the form of forced labour, extractivism, bordering practices and racial capitalism. This is the Caribbean imaginary or, as Édouard Glissant theorised it, a "poetics of relation." Relationality compels us to think beyond the insular and invites us towards archipelagic thinking. As Glissant put it, the entire world is archipelagised and thereby related.[21]
How are the bodies that accommodate bullet fragments in India (see Partha Sengupta's photo essay), Iran (in Behzad Khosravi Noori's essay) and Sweden (the last essay by myself) related to each other? The shotgun pellets in the bodies of the poor people at the Indian-Bangladesh border, or the political dissidents in Iran, and racialized others in Sweden display a dark conjuncture of ethnic nationalism, religious fundamentalism, and white racism.

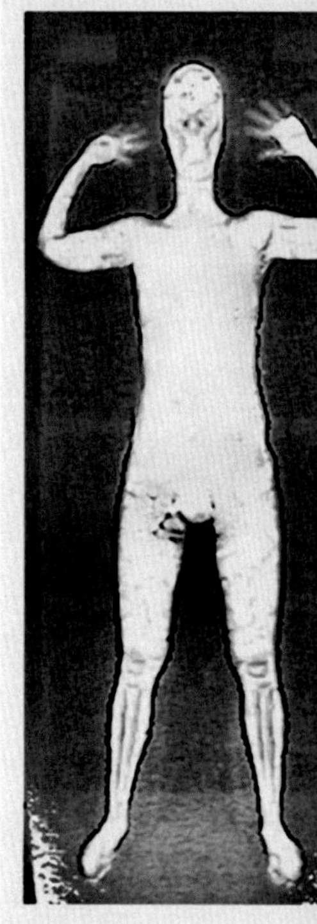

Furthermore, the X-ray archive shows how civilisation and ruination are dialectically interrelated. As a progression in medical science, X-ray technology was fashioned to maintain and save life. However, as cases in this book show, it has been a device of ruination as well. Ruination of life chances, of bodies, of future imaginaries. Ruination is a kind of biopolitics. Whose life is protected against decay? Whose life is exposed to premature decay? Whose body is cared for and its decay delayed? And whose body is abandoned to accelerated decay? Progress and ruination are dialectically related. While on the one hand they negate each other, on the other hand they reproduce each other. As in Javad Abbasi Tavallali's report (this book) on the cruel working conditions for labourers in the oil and gas industry in South Iran, X-ray technology exposes them to cancer-causing radiation. At the same time, X-ray is used as part of health control and detecting health problems (see Yousif M. Qasmiyeh's essay).

[21] Édouard Glissant, *Poetics of Relation* (Ann Arbor, MI: University of Michigan Press, 1997).

However, while X-ray technology has been used as a tool to dominate and to oppress, as some chapters in the book show, it can also be used as a tool to challenge power, through visualization and verbalization of what has been unseen and unspoken in the official archives, histories, and narratives. While a colonial, state-centric, racist, sexist, or racial capitalist archive uses X-ray technology to document, monitor, classify, and ultimately to crystallize otherness. A counter-archive, thus, is inherently an archive from below, which sees X-ray technology as an openness, a potential to expose the violence that is otherwise concealed (see the contributions by Sengupta and Khosravi Noori in this volume). It is in line with what Browne calls "dark sousveillance" as "anti-surveillance, counter-surveillance

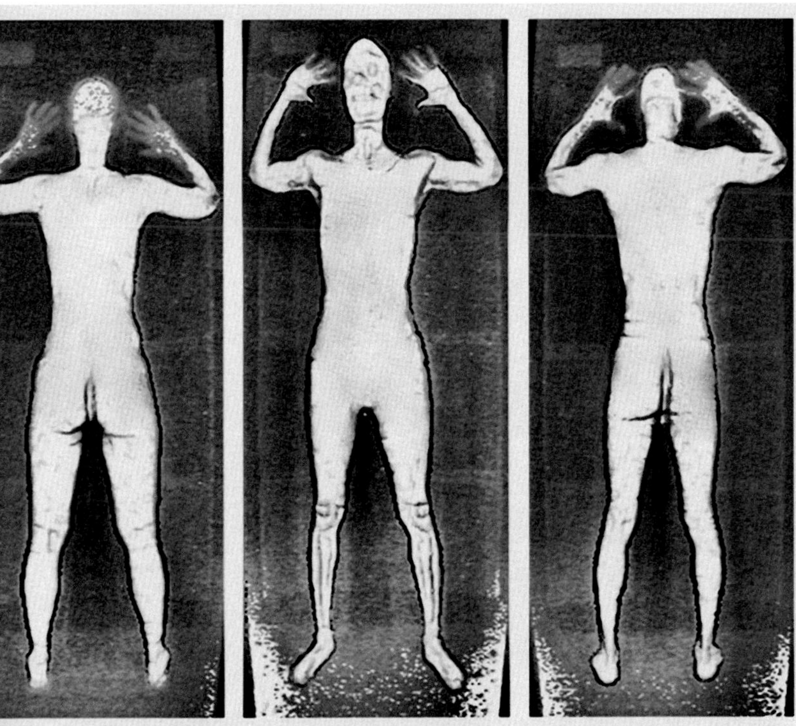

Fig. 6 — Backscatter x ray. Courtesy of Transportation Security Administration.

and other forms of freedom practices."[22] An archive like this one invites us to employ a counter-method. While X-ray is used to regulate the mobility of a colonized people, a counter-method is gaining skills to see through the coloniser; to predict and to evade his power; to deceive him through becoming opaque (see Yara Sharif's essay).

[22] Browne, *Dark Matters*, 21.

X-ray photos are left to interpretation and discretion of the 'experts.' Thus, it becomes a question of power (see Jonathan Krämer's and Reza Hussaini's essays). While X-ray has been used as way to extract 'the truth' (see Françoise Vergès' essay and Paolo Favero's contribution), we can see how X-ray technology has been part of construction of the 'truth' itself (see Ella Hillström and Ayodamola Tanimowo Okunseinde's essay in this volume).

The archive in which sovereign violence is dehistoricised and is thereby naturalized, is an archive from above. An archive from above imposes its own regime of visibility, i.e. how "subjects, subjectivities, and activities have been made visible from above." An archive from below aims "to build up new relations to knowledges and concepts that are not otherwise articulated in the processes of knowledge production."[23]

The X-ray archive from below is an attempt in line with Saidiya Hartman's "critical fabulation," which is an attempt to narrate a certain impossibility and to understand history when all you have are the archives that are constructed against you.[24]

An archive from below refuses and defies how X-ray photos have been produced, understood and used: in order to petrify the Otherness of the bodies of the colonized, indigenous people, the poor, refugees, the racialised and sexualized Other. This refusal is a way to delve into the "incomplete past," searching for stories that tell what has happened or is currently happening.[25]

The contributions to this book make a constellation, which aims to present a diagnosis of the present. *Speculating the impossible* is an attempt to turn the silenced voices in and by the archive audible.

> This silence in the archive in combination with the robustness of the fort or barracoon, not as a holding cell or space of confinement but as an episteme.[26]

[23] Mahmoud Keshavarz, and Shahram Khosravi, eds., *Seeing Like a Smuggler: Borders from Below* (Pluto Press, 2022), 3, 11.

[24] Saidiya Hartman, "Venus in Two Acts", *Small Axe: A Caribbean Journal of Criticism* 12, no. 2 (1 June 2008), 1–14.

[25] Gordon, *Ghostly Matters*.

[26] Hartman, "Venus in Two Acts", 3.

Following Tina Campt's suggestion, we have to "listen to" rather than "look at" images.[27] Campt invites us not only to see but also to watch and to listen to images. She uses the term "haptic temporalities" to show us the multi-layered and complex ways an image signifies through vision, touch, and sound. X-ray images touch us, either to remember or to forget. Like vibrations. We are touched by the vibrations and rhythms. This is the haptic aspect of the X-ray photos. X-ray photos are perceived through several senses and not only through the visual one.

Let's enter the archive from below and see what we hear.

[27] Tina M. Campt, *Listening to Images* (Durham, NC: Duke University Press, 2017).

References

Azoulay, Ariella. "Archive". *Political Concepts, a Critical Lexicon*, 21 July 2017. https://www.politicalconcepts.org/archive-ariella-azoulay/. Accessed 30 September 2023.

Bavli, Itai, and David S. Jones. "Race Correction and the X-Ray Machine—The Controversy over Increased Radiation Doses for Black Americans in 1968". *New England Journal of Medicine* 387, no. 10 (8 September 2022), 947–52. https://doi.org/10.1056/NEJMms2206281.

Benedict, Susan, and Jane M. Georges. "Nurses and the Sterilization Experiments of Auschwitz: A Postmodernist Perspective". *Nursing Inquiry* 13, no. 4 (2006), 277–88. https://doi.org/10.1111/j.1440-1800.2006.00330.x.

Benjamin, Walter. *Walter Benjamin: Selected Writings, 4: 1938–1940.* Edited by Howard Eiland and Michael W. Jennings. Cambridge, MA: Belknap Press, 2006, 389–400.

Browne, Simone. *Dark Matters: On the Surveillance of Blackness*. Durham, NC: Duke University Press, 2015.

Campt, Tina M. *Listening to Images*. Durham, NC: Duke University Press, 2017.

Cartwright, Lisa. *Screening the Body: Tracing Medicine's Visual Culture*. NED-New edition. Minneapolis, MN: University of Minnesota Press, 1995.

Colomina, Beatriz. *X-Ray Architecture*. Zürich: Lars Müller Publishers, 2019.

Derrida, Jacques. *Archive Fever: A Freudian Impression*. Translated by Eric Prenowitz. Religion and Postmodernism. Chicago, IL: University of Chicago Press, 1996.

Dijck, José van. *The Transparent Body: A Cultural Analysis of Medical Imaging*. Seattle, WA: University of Washington Press, 2005.

Fanon, Frantz. *Black Skin, White Masks*. New York: Grove Press, 1967.

Glissant, Édouard. *Poetics of Relation*. Ann Arbor, MI: University of Michigan Press, 1997.

Gordon, Avery F. *Ghostly Matters: Haunting and the Sociological Imagination*. Minneapolis, MN: University of Minnesota Press, 1997.

Hartman, Saidiya. "Venus in Two Acts". *Small Axe: A Caribbean Journal of Criticism* 12, no. 2 (1 June 2008), 1–14. https://doi.org/10.1215/-12-2-1.

Keshavarz, Mahmoud, and Shahram Khosravi, eds. *Seeing Like a Smuggler: Borders from Below*. Pluto Press, 2022.

Khosravi, Shahram. "Bordered Imagination". In *Infrastructural Love*. Basel: Birkhäuser Verlag GmbH, 2022.

Mann, Thomas. *The Magic Mountain*. United Kingdom: Vintage Books, 1969.

Musso, Sandrine. "The Truth of the Body as Controversial Evidence: An Investigation into Age Assessments of Migrant Minors in France". In *Waiting and the Temporalities of Irregular Migration*. London, New York: Routledge, 2020.

Peña, Carolyn Thomas de la. "'Bleaching the Ethiopians': Desegregating Race and Technology through Early X-Ray Experiments". *Technology and Culture* 47, no. 1 (2006), 27–55. https://doi.org/10.1353/tech.2006.0091.

Saunders, Frances Stonor. 'Where on Earth Are You?' *London Review of Books*, 3 March 2016, 7–12. https://www.lrb.co.uk/the-paper/v38/n05/frances-stonor-saunders/where-on-earth-are-you. Accessed 30 September 2023.

Sengoopta, Chandak. *Imprint of the Raj: How Fingerprinting Was Born in Colonial India*. London: Pan, 2004.

Smith, Evan, and Marinella Marmo. "Uncovering the 'Virginity Testing' Controversy in the National Archives: The Intersectionality of Discrimination in British Immigration History". *Gender & History* 23, no. 1 (2011), 147–65. https://doi.org/10.1111/j.1468-0424.2010.01623.x.

Waldby, Catherine. "Medical Imaging: The Biopolitics of Visibility". *Health* 2, no. 3 (1 July 1998), 372–84. https://doi.org/10.1177/136345939800200306.

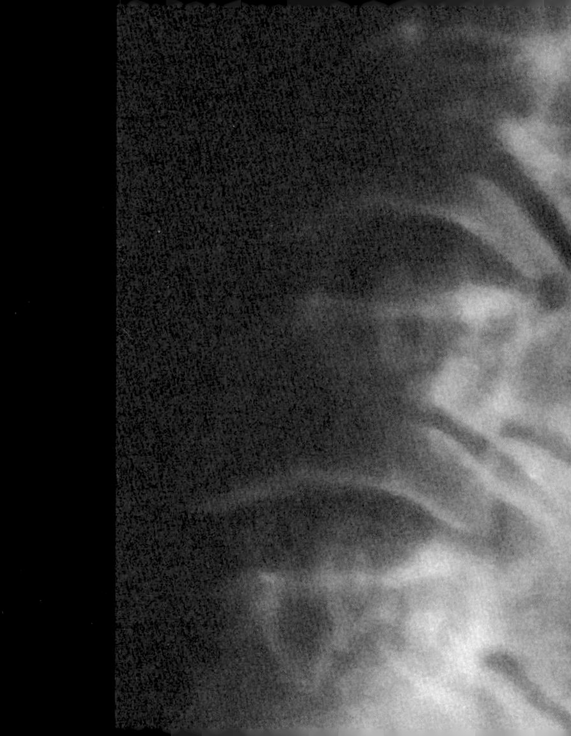

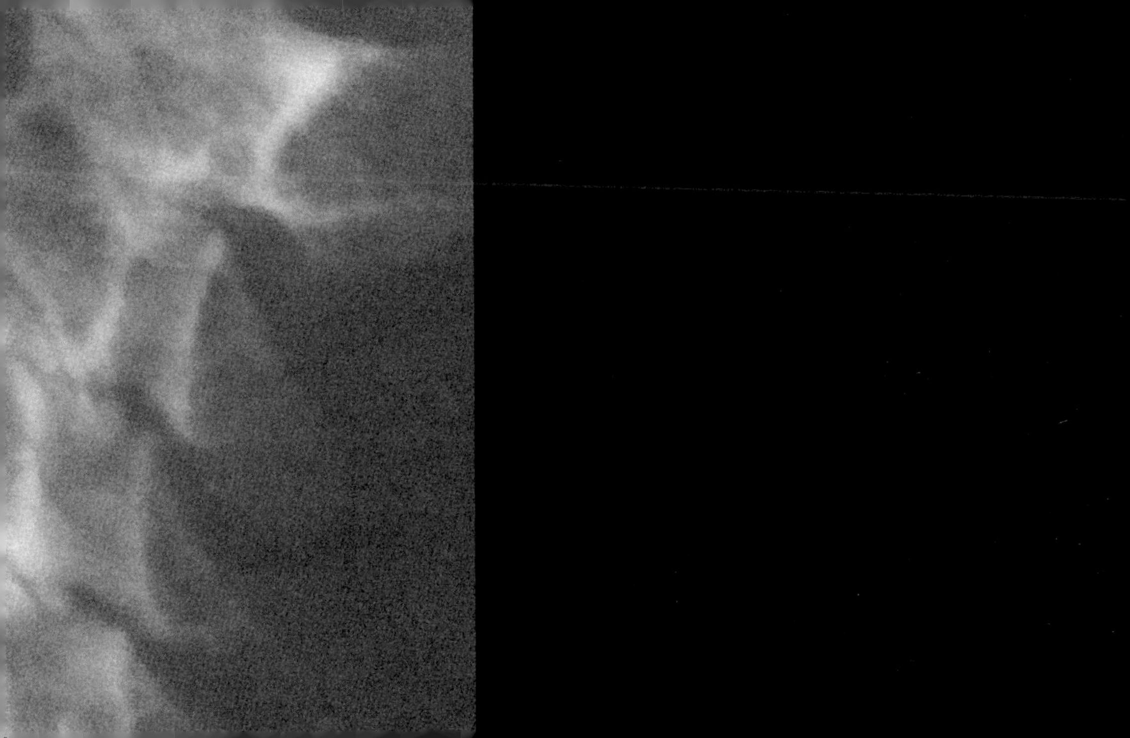

Françoise Vergès

The Racialized X-ray Code

I am standing in a queue in an international airport, it does not matter much where because what happens next has become standard procedure in many countries: I must be submitted to an X-ray machine to check if I am illegally carrying arms, drugs, diamonds etc. I wait patiently. Finally, an airport security employee gestures me to move on. I enter a passenger scanner. I know what to do. I stand facing a silhouette, legs apart, and arms above my shoulders as if I were surrendering to the police. I must not move. The silhouette is shaped like a male body, no waist, no hips. It evokes the silhouette I see in movies during police target practice. I know that I can ask for a pat-down search instead of going through the body screening, but I never do, not only because nothing encourages me to do so but also, I must admit that I tell myself, "Why bother?" It has become part of the travelling routine and a condition of 'freedom of circulation' which we know is not universal.

Wanting to know how the State justifies that process, I look first at the United States Environmental Protection Agency website, which tells me that

> in airports, metal detectors and millimetre wave machines use low energy, non-ionizing radiation to send energy across scanned surfaces. The energy that bounces back from the scanned surface will show the objects that are present, or it can generate an image that TSA [Transportation Security Administration] agents can use to show items that may need more investigation ...

and that

> Backscatter passenger scanners are used to detect threats such as weapons or explosives that a person could be carrying under their clothing. Backscatter machines use very low energy x-rays that are reflected back to the machine itself. Generally, the amount of radiation received from a backscatter machine equals the amount of cosmic radiation received during two minutes of flight and the risk of health effects is very, very low.[1]

I look at other sites and the answer is the same: not only it is safe for myself, but, by doing it, I am making the airport and my flight safe. I am not to worry. It is for *safety* and *security*. By allowing my body to be submitted to rigorous scanning and policing, I demonstrate my willingness to contribute to collective safety and my adherence to the weapons of the war on terror: body scan, pat-down search, biometric passports, visas... Such is the regime of discipline, surveillance and punishment in the neoliberal, imperialist and racist order.

The New Jim Code

The invention of X-ray imaging in the 19th century opened an entirely new field of visibility. From then on, the inside of a living human body would hold no mystery. In the hands of doctors, cops, prison administrators or the mining industry, X-ray became both a tool of prevention against diseases and a tool to control racialized, sexualized and gendered bodies and very often both. As Shahram Khosravi argues in the introduction, "As a medical technology, the X-ray was used not only to reinforce racial imaginaries about the Black body but also has been used against it" (this volume). Advances in technology have perfected racial bias. Hence MIT-scientists, investigating racial bias in the use of medical images in both private and public datasets by Artificial Intelligence (AI), found that AI could

[1] United States Environmental Protection Agency, "Radiation and Airport Security Scanning", *Overviews and Factsheets*, 15 August 2017. https://www.epa.gov/radtown/radiation-and-airport-security-scanning. Accessed 10 October 2023.

> accurately predict [the] self-reported race of patients from medical images alone. Using imaging data of chest X-rays, limb X-rays, chest CT scans, and mammograms, the team trained a deep learning model to identify race as white, Black, or Asian—even though the images themselves contained no explicit mention of the patient's race.[2]

As sociologist Ruha Benjamin has shown, racism pervades technology and other knowledge production, and their inventions worsen inequalities for already marginalized people.[3] Though they claim the objectivity and neutrality of their work, its unbiased and objective methods, people working in technology write racism into code, even if unknowingly. They produce what Benjamin calls a "New Jim Code." Computational racism reproduces, reinforces, and speeds up racial, class and gender inequities. Not only, she writes, "Racial codes are born of, and facilitate, social control," but "... once something, or someone, is coded, this can be hard to change."[4]

X-ray and Anti-Migrant Policies

Under-age asylum seekers are required to take X-ray tests of their wrist bones or teeth to determine if they are really minors and thus receive the protection granted to unaccompanied minors by international and European laws, or if they can be expelled. Once the X-ray of the migrant's left hand and wrist is taken, it is compared with plates that track the ossification progress of children in six-month increments. The standard is the *Radiographic Atlas of Skeletal Development of the Hand and Wrist* first published in 1950 by William "Bill" Greulich (1899–1986), a physical anthropologist and anatomist and (Sarah) Idell Pyle (1895–1987), a Research Associate at the Departments of Anatomy at Western Reserve University and Stanford University School of Medicine, neither author

[2] Rachel Gordon, "Artificial Intelligence Predicts Patients' Race from Their Medical Images", *MIT News / Massachusetts Institute of Technology*, 20 May 2022. https://news.mit.edu/2022/artificial-intelligence-predicts-patients-race-from-medical-images-0520. Accessed 10 October 2023.

[3] Ruha Benjamin, *Race After Technology: Abolitionist Tools for the New Jim Code* (London: Polity, 2019).

[4] Benjamin, 6.

being a clinician.[5] Reprinted multiple times, it is still used by many departments of radiology around the world and remains an important reference in forensic fields. The material in the Atlas was provided by the *Brush Foundation Study of Human Growth and Development*, established in 1929 at Western Reserve University School of Medicine. It collected large amounts of data on skeletal development by X-rays of its enrolled subjects, some of whom were first radiographed at the age of three months. Approximately 800 children from white affluent families in Cleveland, Ohio, were X-rayed for the study by 1930.[6] In other words, white children who were well fed and had access to public health constituted the standard.

Though scientists have demonstrated that the 'Greulich and Pyle standard' (described above) is imprecise and should be used with caution, tests of chronological age remain the tool used by laws and policies that rely on age as a marker or boundary for granting asylum. In France, where, as in other European states, the government has pushed for a growing number of anti-migrant and anti-refugee laws, young migrants have to prove that they are minors in order to be taken into care by the child welfare services (in France the *ASE—Aide sociale à l'enfance*) run by the departmental administrations. Otherwise, they are abandoned while at the same time being under the threat of constant deportation. After examination of the young migrants' identity documents and a social assessment that can take from five days to several weeks, they are often asked to undergo bone testing. "Bone testing proves nothing. You might as well look in coffee grounds or a crystal ball, it'll be just as reliable!"[7] insists Clémentine Bret, referent for minors at risk at Médecins du Monde France.

The unreliability of the test belongs to the logic of a medical research that has stubbornly refused to look at its racial history. The white, healthy, valid, (male) body remains the universal standard. But should we fight for more reliable tests,

[5] Daniel J. Bell, "Radiographic Atlas of Skeletal Development of the Hand and Wrist | Radiology Reference Article | Radiopaedia.Org", *Radiopaedia*, 7 October 2019. https://doi.org/10.53347/rID-71468. Accessed 10 July 2023.

[6] Cf. Bell.

[7] Oriane Mollaret, "Mineurs isolés étrangers: les tests osseux en question", *POLITIS*, 12 March 2019. "Les tests osseux ne prouvent rien. Autant regarder dans du marc de café ou une boule de cristal, ce sera aussi fiable!" Translation author's own.

or for the end of any kind of policy that seeks to hinder circulation across borders, to criminalize migrants and refugees, to institute a constant climate of suspicion, denunciation and push back?

The 'truth' that X-ray tests are seeking is the truth of racial thinking and State surveillance: feeding anxiety, fear, obedience and humiliation to maintain bourgeois order.

References

Bell, Daniel J. "Radiographic Atlas of Skeletal Development of the Hand and Wrist | Radiology Reference Article | Radiopaedia.Org". *Radiopaedia..* https://doi.org/10.53347/rID-71468. Accessed 10 July 2023.

Benjamin, Ruha. *Race After Technology: Abolitionist Tools for the New Jim Code.* London: Polity, 2019.

Gordon, Rachel. "Artificial Intelligence Predicts Patients' Race from Their Medical Images". *MIT News | Massachusetts Institute of Technology,* 20 May 2022. https://news.mit.edu/2022/artificial-intelligence-predicts-patients-race-from-medical-images-0520. Accessed 10 October 2023.

Mollaret, Oriane. "Mineurs isolés étrangers: les tests osseux en question". *POLITIS*, 12 March 2019. https://www.politis.fr/articles/2019/03/mineurs-isoles-etrangers-les-tests-osseux-en-question-40121/. Accessed 10 October 2023.

United States Environmental Protection Agency. "Radiation and Airport Security Scanning". *Overviews and Factsheets*, 15 August 2017. https://www.epa.gov/radtown/radiation-and-airport-security-scanning. Accessed 10 October 2023.

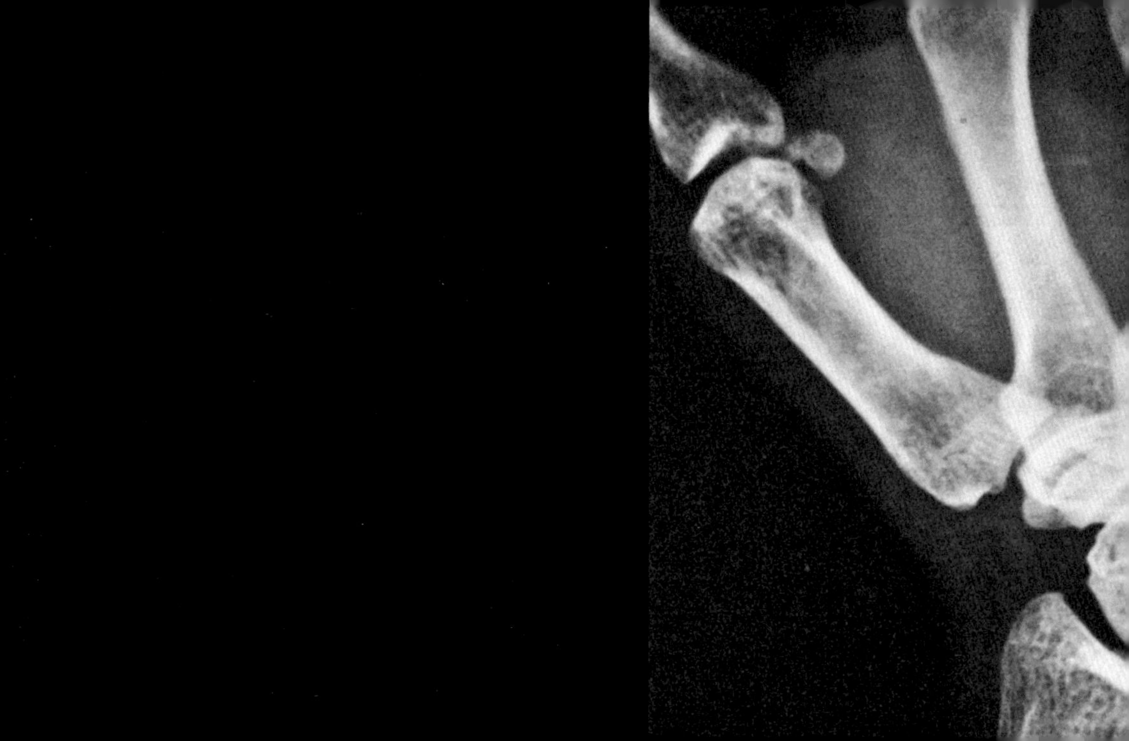

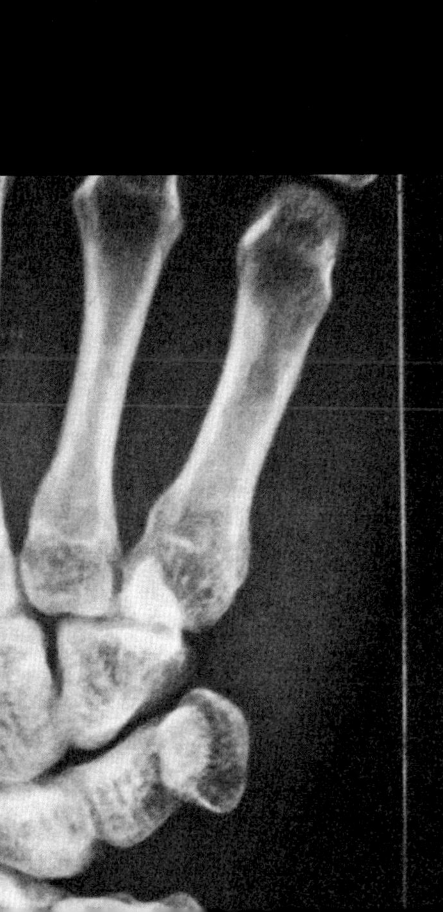

Jonathan Krämer

Moving Shades: X-ray Mobility in Southeast Asian Labour Migration

X-rays in the Migration Clinic

Late one afternoon I found myself sitting on a flimsy plastic stool in a medical clinic on Lombok, a small island in Indonesia. I was doing research on the medical examinations Indonesian migrant workers have to undergo before going abroad, and since many migrants come from Lombok, the clinics there were just the right place. That afternoon I was listening to Ahmad as he listed all the different afflictions he could diagnose with the ageing X-ray machine in front of us: bronchitis, TB, pleural effusion, or abnormal Cardiothoracic ratio. TB and effusion, he emphasized, were the most critical diagnoses. Ahmad was a radiographer, his job being primarily operational, so technically he didn't diagnose. He would take the chest X-rays and then send off the images to a radiologist trained in the art of interpreting them. In fact, the images would be sent to one radiologist in Indonesia, the country of origin, and one in Malaysia, the country of destination. Still, Ahmad had become something of an expert himself over the years, so he could tell the signs of disease from the black and white shadows on the computer screen that form chest X-ray images.

The migration clinic is a key site for Indonesian labour migrants. They apply for jobs in factories, plantations, the service sector and as domestic workers in countries across Southeast Asia, mainly Malaysia, Taiwan and Hong Kong, and on the Arabian Peninsula. In all these countries, medical examinations are compulsory at different stages during the application and renewal process for employment visas.[1] The stakes are high for the migrants examined. An 'unfit' result can mean the end of the migration process even before they have departed or worse, deportation once they have reached their destination. These examinations must comply with standards set by the governments of the destination countries, which determine what is to be examined and how, and what diagnoses result in a failure to pass the examination. Moreover, in most cases, acceptable medical certificates can only be issued by institutions accredited by the destination country's health authorities and examination results might even need to be submitted directly to immigration authorities through specific online systems. A migration clinic then, is a specialized medical institution focusing exclusively, or almost exclusively, on conducting these medical examinations of prospective labour migrants. As such, it is at once a key component of what Biao Xiang and Johan Lindquist have called "migration infrastructure,"[2] facilitating the movement of migrants abroad and part of a de-localized regime of border control that displaces the process of selecting admissible individuals from the actual border checkpoint.[3]

Images Collected

The chest X-ray is one of the main technologies used in the migration clinic, with every migrant having to undergo a radiological examination of their chest as part of the standard examination procedure. In the case of West Malaysia, one of the main destinations for Indonesian migrant workers, X-ray images must be submitted digitally through an online system. Each image is assessed twice, once by an Indonesian radiologist working at the migration clinic and once by a Malaysian radiologist, who accesses it through the online system. As a representation of the migrant's body, the X-ray image travels far, while the migrant remains in place, awaiting results and

[1] Cf., for example, Aaron Parkhurst, "Blood, Lungs, and Passports", *Medical Materialities* (London: Routledge, 2019), 85–97.

[2] Biao Xiang, and Johan Lindquist, "Migration Infrastructure", *International Migration Review* 48, no. 1_supplement (1 September 2014): 122–48.

[3] Mark B. Salter, "The Global Visa Regime and the Political Technologies of the International Self: Borders, Bodies, Biopolitics", *Alternatives* 31, no. 2 (1 April 2006): 167–89.

decisions. In doing so, it is decoupled from the migrant, who will most likely never get a chance to see it unless the radiographer permits them a quick glance at his computer screen right after taking the image. The medical examination, mostly a group affair, is too tightly structured to give migrants the time and space to engage with the X-ray images of their own bodies.

Of course, medical examinations in the migration clinic involve other representations of migrant bodies as well. There are, for example, the blood and urine samples used to test for infectious disease, with a few drops taken to represent the entirety of the body's fluid. Or the blood pressure taken during the physical examination, where the momentary systolic and diastolic blood pressures are extrapolated to represent a generalized condition of the migrant's body. Each of these medical representations is interpreted in the process of examination and the results of these interpretative acts are duly noted on the form, which is subsequently translated into a medical certificate. On the medical certificate, the migrant's body is reduced to a range of interpretations that lend meaning to the medical representations produced during the examination. There may even be interpretative guidelines revealing what the ideal result, the reference value, should be. While for many medical categories in the certificate the diagnostic options are binary, testing positive or negative, the radiological exam allows for a more open-ended range of diagnoses, including common conditions like bronchitis and rarer, potentially more serious diseases, such as tuberculosis. Ultimately, however, even these diagnoses are distilled into binary results, too, with the X-ray image being described as either 'normal' or 'abnormal' and finally the entire medical exam resulting in an overall judgement of 'fit' or 'unfit.' It is this judgement that really matters for the migration process, as 'fit' means that the prospective migrant can receive a visa and leave for work abroad, whereas 'unfit' means that they will not be issued a visa until the diagnosed condition is cured, effectively stalling the process.

Yet the difference between 'normal' and 'abnormal' or pathological bodily conditions is much less clear cut than its terminology suggests. Georges Canguilhem famously argued that distinguishing between the normal and the pathological body is essentially a matter of normative judgement rather than objective fact, so

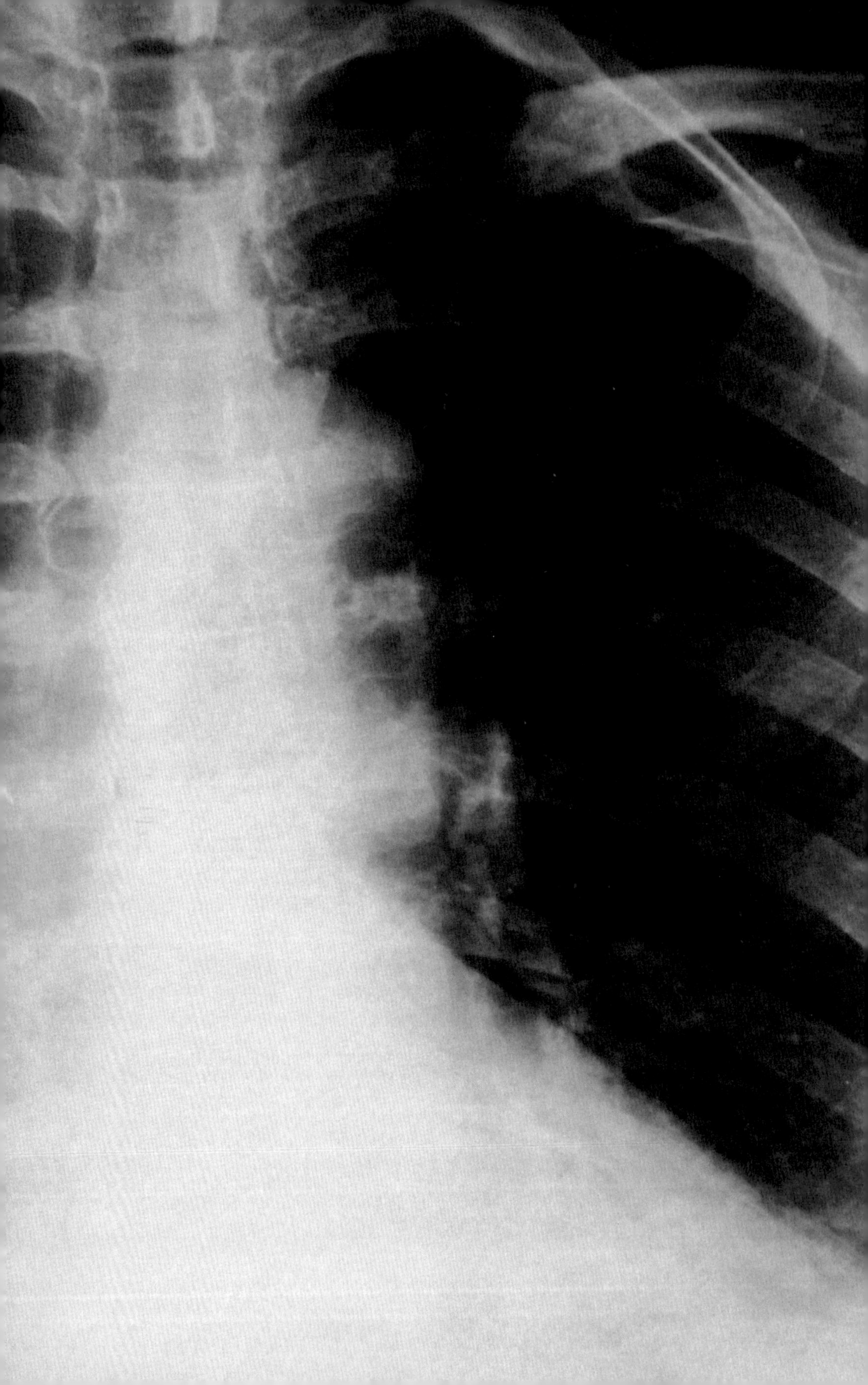

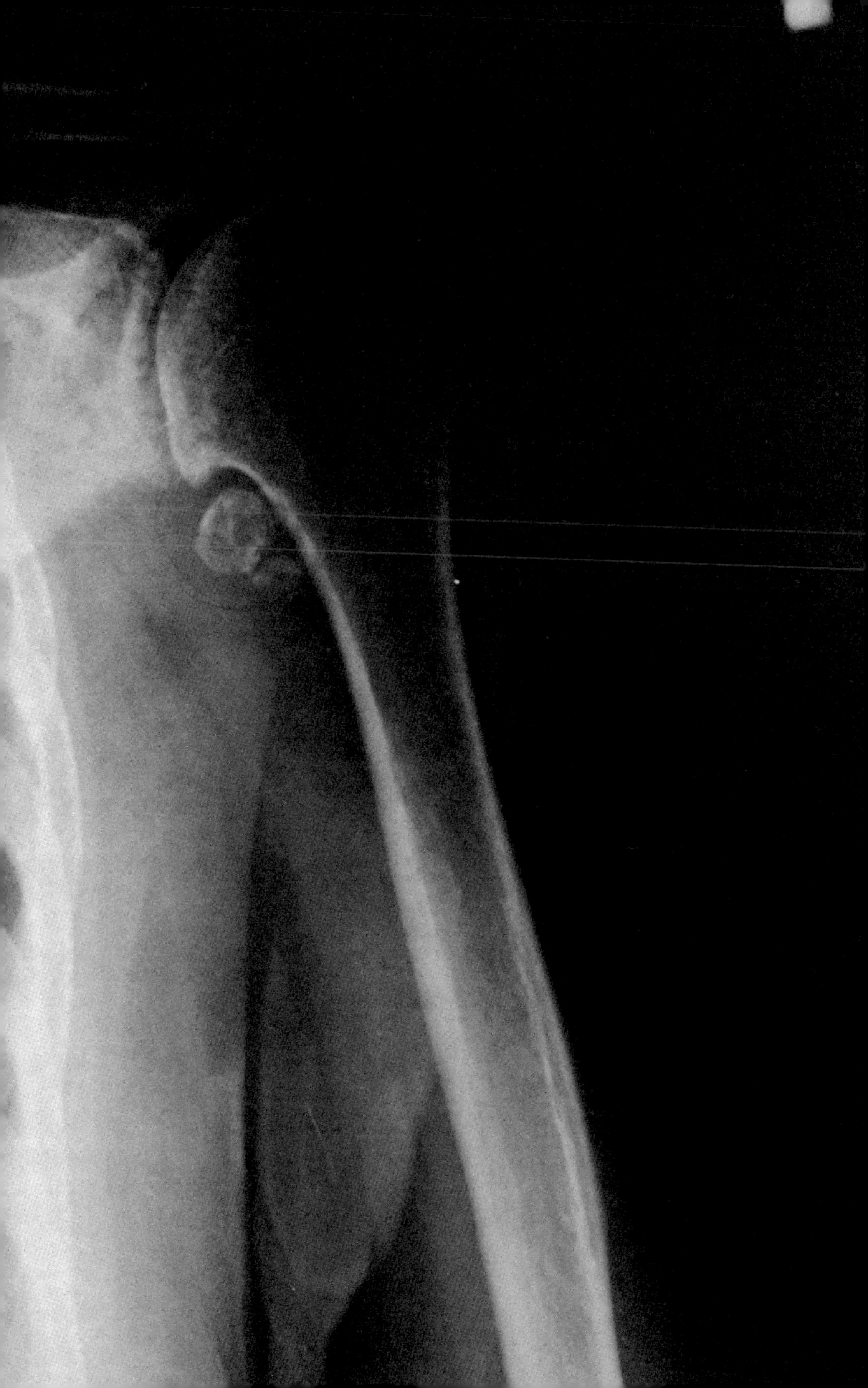

that the threshold dividing them necessarily remains unsettled.[4] What is more, the X-ray image itself is, ironically, a particularly opaque representation of the body. As Ehsan Samei writes in his introduction to a special issue of the *Journal of the American College of Radiology* devoted to the perception of medical images: "Medical images do not clearly portray reality. Consequently, certainty is provided not by an image itself but by the interpretation of the image."[5] In practice however, this interpretation, the "reading of the image" as it is commonly called by health professionals in the migration clinic, often fails to create certainty and in fact may lead to disagreement among doctors. Thus, studies on chest X-rays for tuberculosis screening, for example, suggest a variability in image interpretation of about thirty-three percent between observers, which means that in a third of cases, different radiologists will reach different interpretations of the same image.[6] Even more strikingly perhaps, are findings that indicate such inconsistencies in the interpretations of one observer in twenty percent of cases.[7] In other words, in a fifth of cases a radiologist will interpret the same image differently if shown it multiple times.

Images Rejected

"Aduh," he exclaimed, "aduh, aduh, aduh." I was chatting with a group of migrants, when Susilo suddenly jumped up from his desk, where he had been staring intensely at a computer screen and walked across the spacious office shaking his head. *Aduh* is the Indonesian equivalent to "Oh" and Susilo was the clinic's head of administration, so clearly something was amiss. I asked what happened and, letting his head hang low, Susilo simply said, "Banyak sekali yang reject"—many got rejected. The actual number was twenty-eight migrants whose medical certificates had been rejected by the Malaysian health authorities because of differing interpretations of their X-ray images and this spelled trouble for Susilo; trouble and a long day at work.

[4] Georges Canguilhem, *The Normal and the Pathological* (New York, NY: Zone Books, 1989).

[5] Ehsan Samei, "Why Medical Image Perception?", *Journal of the American College of Radiology* 3, no. 6 (1 June 2006): 400.

[6] Jerome Groopman, *How Doctors Think* (Boston, MA: Houghton Mifflin, 2008), 181.

[7] Groopman, 181.

The migrants, who were planning to go to Malaysia to work on plantations and in factories, had paid the clinic a lot of money for the exam and most of them had passed pre-examinations to make sure these expenses would not be in vain. The rejection of these medical certificates meant that Susilo would have to field angry calls from the brokers of the migrants affected and negotiate conditions for a re-examination with them. It would cost the clinic a lot of time and, worse perhaps, a lot of money. Susilo left the room still shaking his head and running his hand over his face. Meanwhile the migrant brokers sitting next to me laughed and joked about the scene. This sort of thing happened regularly and as long as their migrants weren't affected it made for good entertainment.

While the indeterminacy of the boundary between the normal and the pathological is laid bare during almost all types of examination at the migration clinic, with brokers and migrants contesting most diagnoses, it becomes most salient in the case of chest X-rays, because of their material form. How would one organize the daily transfer of thousands of blood and urine samples to laboratories in the destination country in a viable manner? Or, even more absurdly, find a way to send abroad samples suitable for the tactile examination of lymph nodes and for reading blood pressure? In the case of radiological examinations, the transfer of material across large distances is quite simple. X-ray images can be easily uploaded to online systems and sent for evaluation to radiologists working halfway across the globe in matters of seconds.

Ultimately, it was the outcome of this transferability of the X-ray image and the system of control and distrust requiring X-ray images to be double-checked in the migrants' destination country, that caused Susilo's distress. That day, twenty-eight X-ray images assessed as normal by the migration clinic's radiologist were rejected by the radiologists in Malaysia, who read the images as showing signs of disease, which translated into the migrants' health status being changed to 'unfit'. There was no option for appeal, the Malaysian radiologists' interpretations were final and carried more weight than those of their Indonesian colleagues. In short, these twenty-eight migrants would not be able to go where their X-ray images were so effortlessly sent. At least not along the documented route.

At this point, the journeys of migrants and those of their documentary representations diverge. The migrants might either repeat the entire medical examination in the hope of being luckier the second time around, change their destination to another country, go undocumented, or simply decide to remain in Indonesia for the time being. Meanwhile, what happens to the X-ray images that were uploaded to the health and immigration authority's submission system is even more opaque. Most likely, they continue to be stored in the cloud for archival purposes, lingering in a sort of heaven, or cemetery if you like, of radiographic representations of migrants, whose attempts at moving abroad to work were either enabled or foiled by the mobility of these images.

Thus, the intersection of migrants' mobility and that of their X-ray images is only temporary. Once it has been interpreted by radiologists across both sides of the border, the X-ray image is no longer of use, other than perhaps as proof underpinning the respective readings. Should a migrant who failed the medical examination due to the radiological part of it, wish to try again, they are required to undergo another X-ray exam. A new image is produced, creating a new representation of the migrant's body in the hope that either whatever was the cause for concern in the earlier image would have vanished by now, or that, perhaps more likely, a different radiologist's gaze would yield a more favourable interpretation. In this way, the production and transfer of X-ray images is both implicated in the construction of what Lindquist has termed "channels" of migration, marked by instability and plasticity, and consciously draws on this instability as with each image a new channel is created—perhaps only slightly reconfigured, but new nonetheless.[8]

[8] Cf. Johan Lindquist, "Brokers, Channels, Infrastructure: Moving Migrant Labor in the Indonesian-Malaysian Oil Palm Complex", *Mobilities* 12, no. 2 (4 March 2017): 213–26.

Conclusion

The mobile X-ray image is central to a process that structures and regulates the conditions of migrants' mobility in the name of maintaining public health and protecting the national body of the destination country. As such, it is paradigmatic for the intertwining of measures governing health and immigration at both national and transnational scales. In this context, the X-ray image's particular materiality sets it apart from other medical representations of the body. Digitalized, it can easily be sent across vast distances to be interpreted by medical experts abroad, who have otherwise no way of seeing or touching the body they examine. This movement of X-ray images creates further uncertainty for the migrants whose bodies are radiographically represented. Not only is the number of people involved in evaluating their bodies increasing, but this evaluation is also becoming more and more distanced, reducing the migrants' chances to appeal unfavourable results. Migrants, then, are stuck waiting, unable to cross borders, while X-ray images of their bodies are sent abroad in matters of second. This, perhaps, is the most critical effect of radiography's role at the intersection of border control and public health: as representations of bodies become more mobile, bodies themselves become less so.

References

Canguilhem, Georges. *The Normal and the Pathological*. New York, NY: Zone Books, 1989.

Groopman, Jerome. *How Doctors Think*. Boston, MA: Houghton Mifflin, 2008.

Lindquist, Johan. "Brokers, Channels, Infrastructure: Moving Migrant Labor in the Indonesian-Malaysian Oil Palm Complex". *Mobilities* 12, no. 2 (4 March 2017): 213–26. https://doi.org/10.1080/17450101.2017.1292778.

Parkhurst, Aaron. "Blood, Lungs, and Passports". In *Medical Materialities*. London: Routledge, 2019, 85–97.

Salter, Mark B. "The Global Visa Regime and the Political Technologies of the International Self: Borders, Bodies, Biopolitics". *Alternatives* 31, no. 2 (1 April 2006): 167–89. https://doi.org/10.1177/030437540603100203.

Samei, Ehsan. "Why Medical Image Perception?" *Journal of the American College of Radiology* 3, no. 6 (1 June 2006): 400–401. https://doi.org/10.1016/j.jacr.2006.02.017.

Xiang, Biao, and Johan Lindquist. "Migration Infrastructure". *International Migration Review* 48, no. 1_suppl (1 September 2014): 122–48. https://doi.org/10.1111/imre.12141.

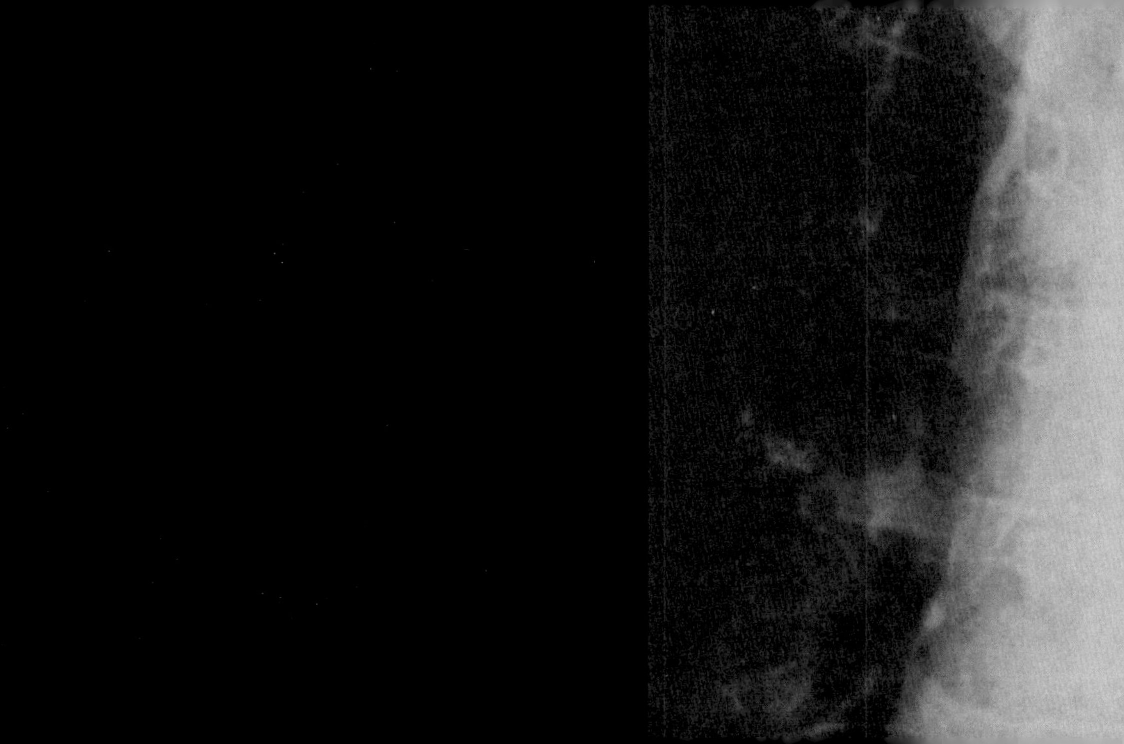

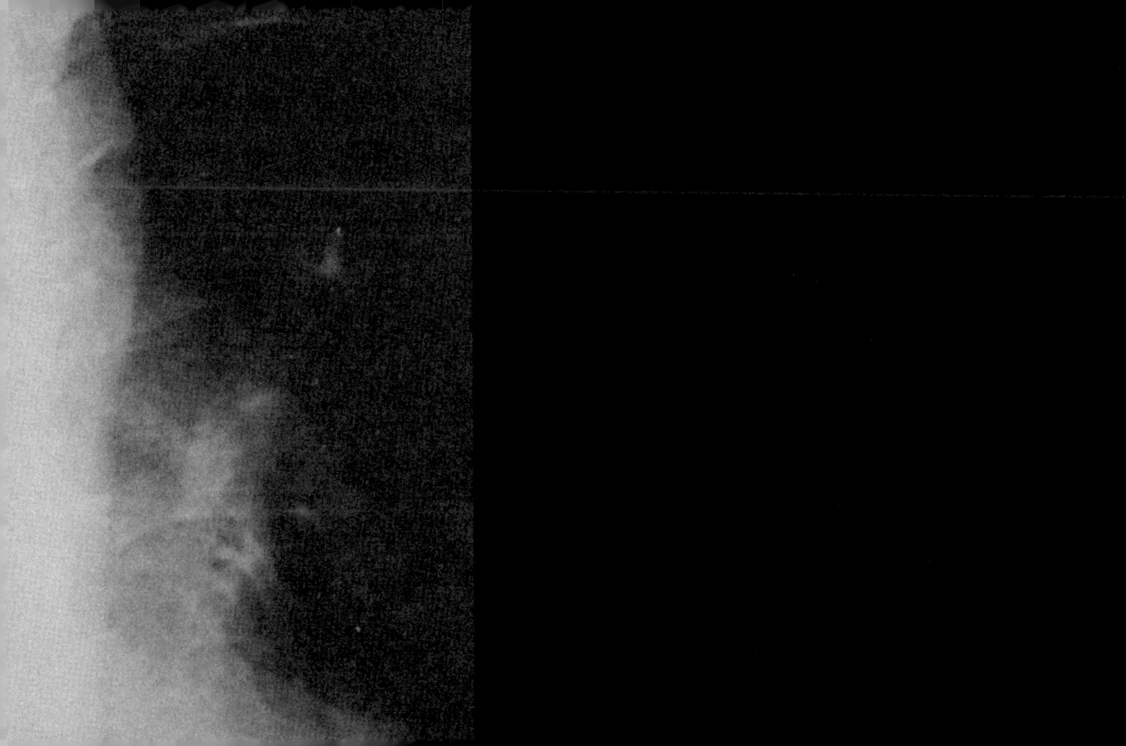

Reza Hussaini

Suspicious Bodies

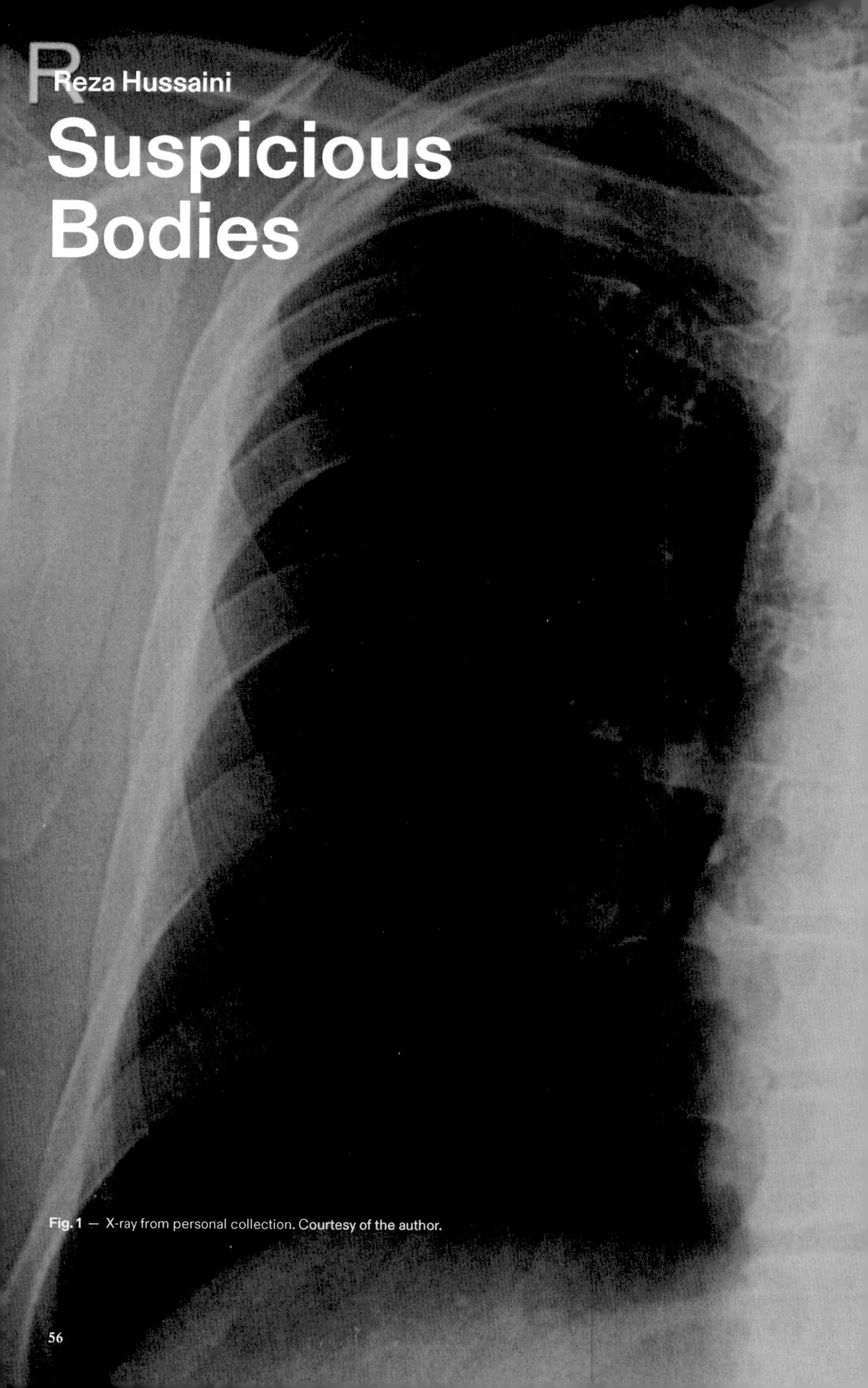

Fig. 1 — X-ray from personal collection. Courtesy of the author.

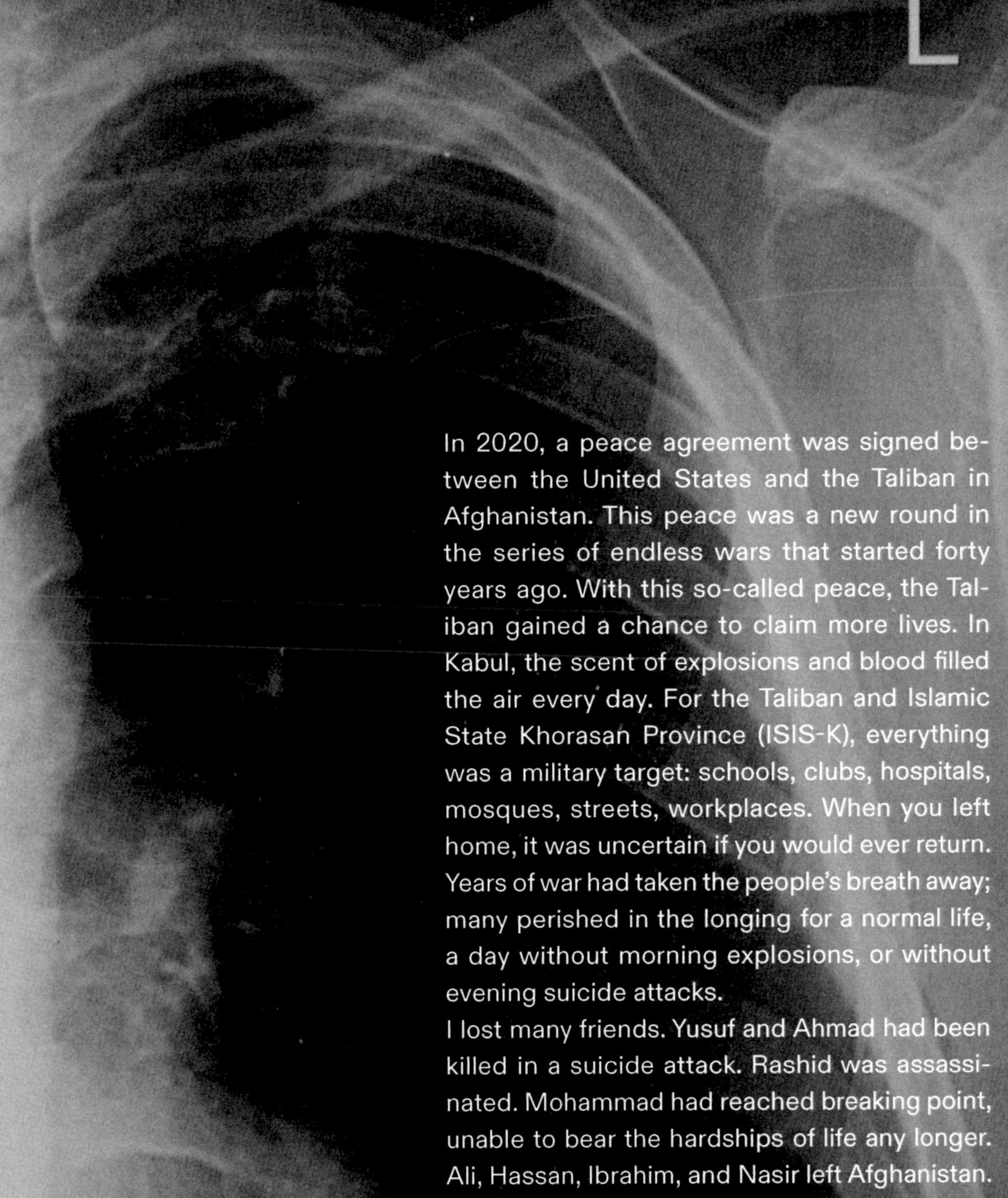

In 2020, a peace agreement was signed between the United States and the Taliban in Afghanistan. This peace was a new round in the series of endless wars that started forty years ago. With this so-called peace, the Taliban gained a chance to claim more lives. In Kabul, the scent of explosions and blood filled the air every day. For the Taliban and Islamic State Khorasan Province (ISIS-K), everything was a military target: schools, clubs, hospitals, mosques, streets, workplaces. When you left home, it was uncertain if you would ever return. Years of war had taken the people's breath away; many perished in the longing for a normal life, a day without morning explosions, or without evening suicide attacks.

I lost many friends. Yusuf and Ahmad had been killed in a suicide attack. Rashid was assassinated. Mohammad had reached breaking point, unable to bear the hardships of life any longer. Ali, Hassan, Ibrahim, and Nasir left Afghanistan. Others were searching for a way out. In restaurants, workplaces, taxis and homes, people talked about 'leaving the country.'

In 2020, I obtained a scholarship for a doctoral program at a university in the United Kingdom. However, the main problem was getting a student visa. The visa application meant providing various forms and documents. To fill out the forms caused a sense of humiliation. They demanded all the details of my personal life: the names of my parents, spouse, children; their dates of birth; their places of residence and details about my travels over the past ten years.

Where did you go?
When did you go?
Why did you go?
When did you come back?

It was an extensive interrogation.

In either peace or wartime, have you ever been involved in, or suspected of involvement in, war crimes, crimes against humanity, or genocide? Have you ever been involved in, supported or encouraged terrorist activities in any country? Have you ever been a member of, or given support to, an organization which has been concerned with terrorism? Have you ever been a member of, or given support to, an organization which is or has been concerned with extremism? Have you, by any means or medium, expressed any extremist views?

I was confused. I did not understand why I was being asked these questions. Why, how and when were these questions included in the visa application form? As far as I remember we had been the *victims* of war. It took me several days to convince myself to complete the form. There was no way to enter the UK without answering those questions on the application form. It seemed to me a kind of confession, a religious ritual that was alien to me.
After filling out the forms, I had to undergo a tuberculosis test. The test was only mandatory for the applicants from Afghanistan, and other 'suspicious' countries such as Iraq or Sudan.
To enter England from a country like Afghanistan, 'confession' forms were not enough; I had to prove that my body was free of all contaminations. The assumption was, the Afghan body was contaminated unless proved otherwise. The test, the evidence of purity, was a ritual, part of a rite of passage. For the test I had to provide a chest X-ray.
The British embassy demanded the test be done by the office of The International Organization for Migration (IOM) in Kabul. I could get X-rayed cheaper and faster in a private clinic in Kabul but it would not be approved. I received an appointment for three weeks later.

On the day of the test, I sat in a taxi, thinking about the morning suicide attacks. I was scared. Explosions and suicide attacks mostly happened on crowded streets in the morning or evening when people were going to or home from work.
The IOM office was in Kolola Pushta. The building was protected by tall concrete blast walls. I had heard that each section of wall weighed around 10 tons and cost between US$500 to US$600. A lucrative business. Kabul was full of these blast walls. In the middle of the tall perimeter wall of the IOM office, there was a large steel entrance door controlled by the first security layer. As I approached, an Afghan soldier asked my name to check if it was on the list. He gave me a mask and asked me to wash my hands according to the health protocols and then disinfect them. He pounded heavily on the armoured door a few times. On the other side, a ten-centimetre metal hatch in the middle of the door slid open, revealing the eyes of a soldier. He looked at me and the heavy door opened and I entered a hall. It was the second stage of the security check. I had to put all my belongings in a basket, which rolled through an X-ray machine and I had to go through a security scanner. It was like being at an airport and it felt as if I was travelling abroad. After that, I reached the reception, where they asked for my passport and an Afghan soldier guided me inside the main building.
After passing the first section, I did not see Afghan soldiers any more. There were foreign armed security guards watching me. They seemed to be Asian. In Afghanistan, I had seen soldiers from various countries: Americans, Europeans, and Australians, but not many from Asian countries.
Once again, I was asked to fill out a form with detailed personal information. Then an employee attached a photo of me to the form and said, "You have to pay US$138 for the test."
Shocked, I asked, "Why is it so expensive? A local clinic is doing it for US$10."
The employee, who seemed irritated, said, "I am just an employee here and I am only doing my job."
I said, "In a country where people earn US$2 a day, this is not fair."
He replied, "Do you know how much the salary of these Filipino soldiers who are responsible for the security here is?"
I said, "Why should I pay the salary of Filipino soldiers?"

Baffled, I went to the bank and paid the fee and returned, repeating the same security procedure, watched by the Filipino guards. After delivering the money receipt, I was given the address of the IOM X-ray testing place. The address was a few streets away. It was a small shopping mall with clothes and shoe shops. Doubtfully, I asked one of the shopkeepers if the IOM clinic was there. He pointed at the end of the mall. In a small room, approximately twenty square meters big, was an X-ray machine. The room was packed with Afghans who needed to do the tuberculosis test. I handed over the IOM letter and the testing procedure began. I was told to take off my shirt and place my naked chest against the machine. The X-ray technician left the room and came back after a few minutes. It was done. He said that the test results would be sent to the IOM office on a CD, and he gave me a copy of the X-ray image.

According to the IOM doctors, I was suspected of having tuberculosis. They detected opacity in some airspaces within the lungs. This was not surprising. The heavy polluted air in Kabul caused by the burning of coal, wood and plastic during the winter damaged lungs. The air in Kabul was so polluted during the winter that we could hardly see a few steps ahead. And during summers, we breathed in dust and dirt.

My visa process was suspended due to the opacities discovered in the X-ray image. Many other applicants had similar problems. We all visited the IOM office for three consecutive days for sputum testing. The samples were sent to Pakistan for cultivation. It took two months to get the result. It was negative and I could prepare myself for the next stage of my journey to the UK. However I did not feel relief but rather diminished, humiliated and abused.

Unlike the IOM building which was impermeable, shielded by concrete blast walls, I was turned into a translucent body. First through the confession in the forms and then by X-ray. The sense of nakedness in front of a powerful, faceless, soulless power was unbearable. The whole process seemed designed to tell me that I was nothing, just a shadow on a slide.

Reza Hussaini

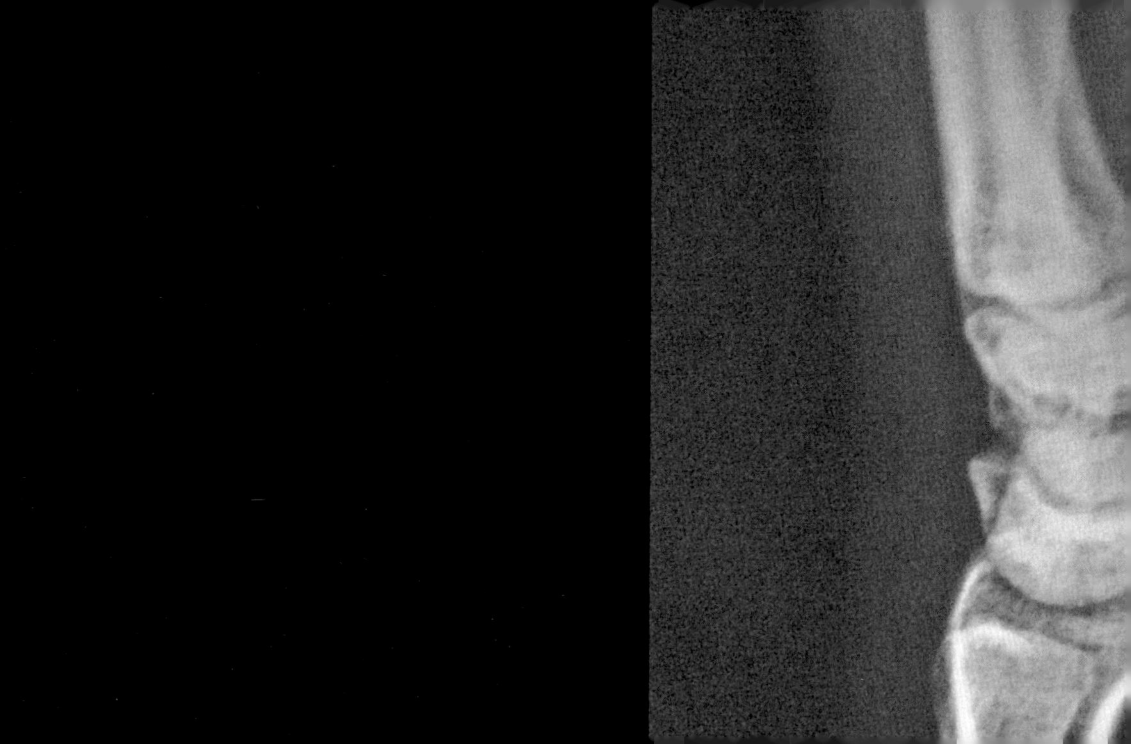

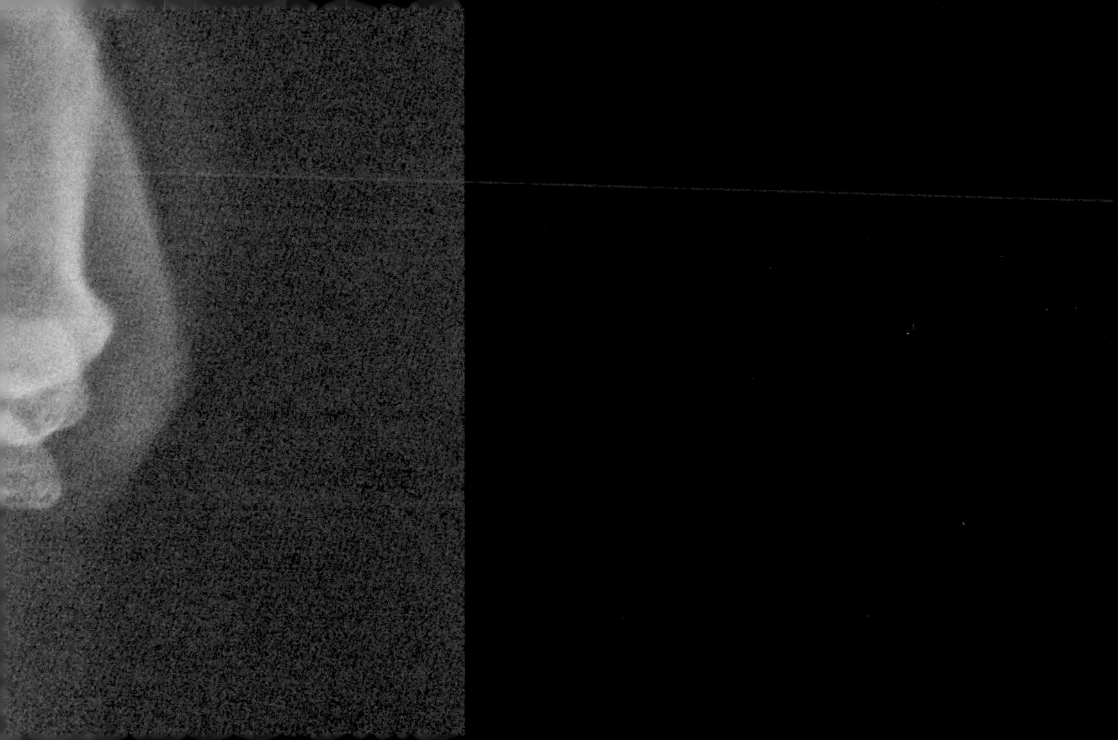

Yousif M. Qasmiyeh

The Ancestral Document

Historicising Pain

Rendered fleshless, the body through X-raying tells their story instead of telling its own. I was X-rayed soon after claiming asylum in the United Kingdom. In Terminal 4 of Heathrow Airport. I recall standing outside a room accompanied, or rather monitored, by an official. I was kept with dozens of other asylum seekers, not in the same room, but in a separate zone, on a different floor. My body was fully palpated, especially the upper part, then I was X-rayed. Not knowing why. Not knowing how it all happened before my eyes and yet without my presence.

On my body, there are ancient scars from burns, vaccinations, and shrapnel finding their home in different crevices. Even though, under body hair, such scars can no longer be easily discerned or located by the untrained eye, they never disappear. That is exactly why they retain their omnipresence even in their temporal absence. With such seeking, and by seeking, I mean seeking the hidden, I invite the seer (as well as the reader) to find my underburied self in time, in old wounds upon which corporeal twigs and branches are scattered. In my mind's eye, I can still see them despite not thinking of their place. I assume that this is seeing as historicising, as a constant search for that which makes me who I am; a refugee who passed through time but also marked by it—the kind of seeing that transcends vision so it would follow its course to past lands.

Such scars are also shared with my siblings for the sheer fact of sharing the camp. As we were growing up in Baddawi camp in Lebanon, we ran competitions on whose scars were the most visible and therefore more credible. For us, as children, the credibility of pain hinged on visibility, a formula which was soon rendered futile.

In the camp, we are X-rayed for us and for them. As my mother often affirmed: It is *them* before *us*. For them so they would be free of that which we harbour; of ailments. For them again so they are eternally the ones who care for us. Infantile beings the refugees are, as they, in the eyes of the many, lack the basic capacity to care for themselves or even to declare *gesturally* that they need help. In this way, the refugee is the only being whose health is girded (and guarded) by his ill-health. We would be in queues, long, infinite lines, starting at school, where instead of learning the six subjects of the day we would learn the art of patience: we would be taken to the camp's clinic for X-rays. As for vaccines, they would happen at school, sparing us the need to cry en route but instead in queues, then shepherded into two zones: one for those whose bodies were by then accustomed to pain and needles, and the other for those whose reticence would at times leave broken needles in the flesh.

X-raying Intentions

Refugees must rummage through their aches as the sound in them is seen as a threatening ability whereas unwellness, despite its ramifications, is what makes the citizen and the official who they are: spectators awaiting the tragic death of a pathetic hero. Not long ago, I had an X-ray of my upper body, now as a naturalised citizen. "Breathe in. Breathe out," said the woman in the hospital. I stayed still afterwards with arms stretched forwards clasping air. An embrace for that void that we benignly call the body. In a brownish envelope in the camp, my father's X-rays were accumulated, all put forth to validate ailments. The clear proof that our illness is universal.

The trove of documents assembled by my father himself, he who fought during different incursions into the camp and lent his fighting hand to other refugee camps in Lebanon. His X-rays narrate his body. His ailments, wounds and scars. They validate that which is hidden to the naked eye. He used to say to the UNRWA[1]

[1] United Nations Relief and Works Agency for Palestine Refugees in the Near East.

people trying to cut our monthly or 3-monthly rations: "See, it's all there, my incapacitated body is there for you all to see." Not only the X-ray but a medical report delineating every defect.
Not long ago, my mother disposed of the once-treasured X-rays. My father's stomach. His spine. His legs. Every limb and more. As soon as he died, in 2020 she said, "they're no longer of use."

The X-ray is an Ancestral Document

For me, as it was for you, thinking now of your body in its grave, thinking of how it is and will be for us all since the X-ray is that deferred death, seeing through us, seeing us as the absolute nakedness, bodies stripped of their flesh so bones can stand in lieu of the living and the dead alike. According to him, what was being X-rayed, and thereby documented, wasn't just his body as proof of ailment but rather the utter futility of the refugee's speech in his attempts to be believed as an ill being. My mother, acting as the intermediary between UNRWA and doctors, would hop from one hospital to another, in the company of a husband who was hardly there physically and mentally, begging those in charge to ensure that those X-rays would be supplemented by rigorous reports, multiply affirming that my father's illness could not be left solely to time and that it was UNRWA's duty to neither reduce nor cut our rations.

Excavating Pain: X-rays as the Refugee Self

Bodies are elsewhere, even though their grasping seems to be always a possibility for us. It is worth its pain—the scale by which an existence is measured is the same scale that denies our worth as human beings. To put it simply, the real pain that we bear is not sufficient as it is not a pain that is believed a priori. It is always subject to suspicion. Look at this double-bind: subject the body to X-rays in order to become in essence too invisible and yet too visible to be appraised by the UN-body. So the X-rays, or more so the X-raying, was never for our own wellness, or for those pains to be cured or medicated once situated. Far from it, the refugee is a body in pain for his own pain and the proof is for the discerning eyes which want to see through, and never see, the human literally.

This X-raying, for me, could be seen as both an authoritarian and authoritative curation of the refugee body insofar as what is being assembled (or in fact dismembered) is not the body proper but an invisible version of it which can only be read by the so-called expert for another expert in the name of humanitarianism.

For me, since I was young, our amassed X-rays were more the afterlife of an aching body, what the eye could see beneath the body, our nakedness in black and white, or in shades of both colours.

Of late, perhaps due to a combination of a cocktail of sedative medications and anti-depressants due to depression, anxiety and thoughts of self-harm, I have continually returned to my father, to his lateness, to his already-happened death but also to mine in the future and how the X-ray I was subjected to in Heathrow Airport as part of my asylum claim is nothing but an inter-generational exposure (father and son) to that eye that seeks the innards of its other.

Yousif M. Qasmiyeh

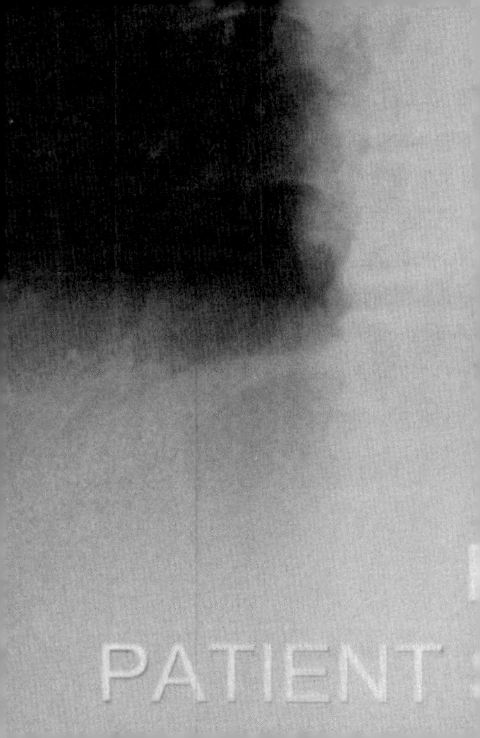
PATIENT

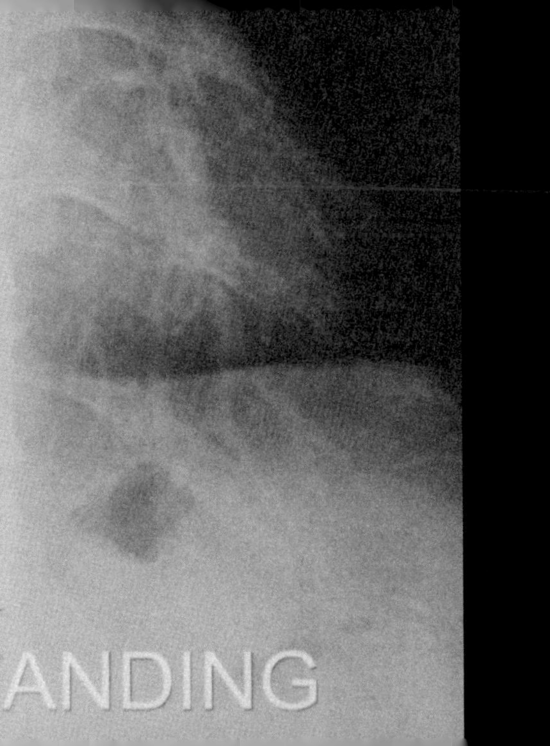
ANDING

Partha Sengupta

The Bloodiest Border. A Photo Essay

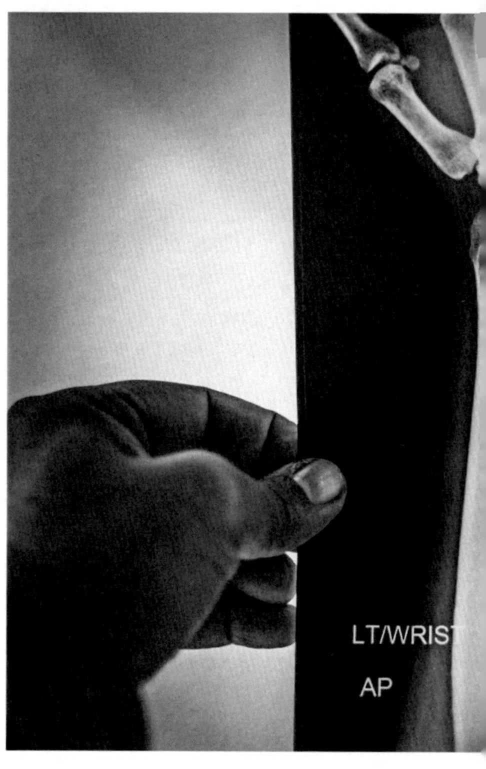

Fig. 1 — Mohor Mondal's broken arm is clearly visible on his X-ray.

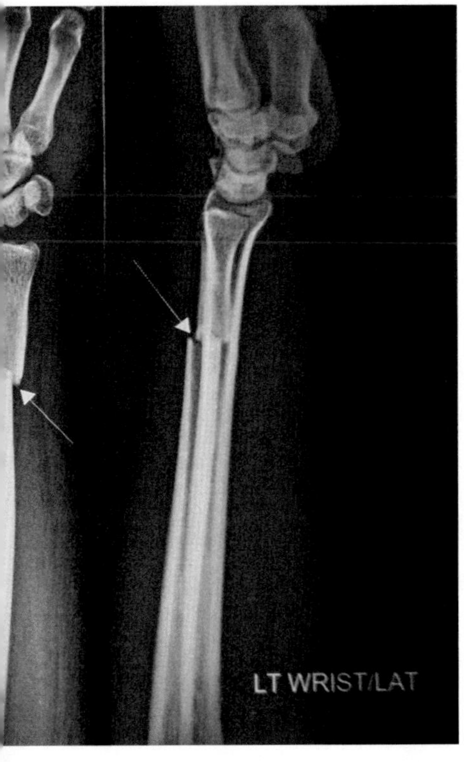

The India-Bangladesh border is 4,096 km long, of which 3,145 km has been fenced, and the rest is porous; the border goes through a river and homes. It is the fifth-longest land border in the world. The border was twice redrawn, upon the partition of India in 1947 after Independence and after the Bangladesh Liberation War in 1971. The border is a colonial construction that has divided the people who have lived in the region for generations, which turned into two separate nation-states. It is not only one of the longest, but the border between the state of West Bengal in India and Bangladesh (2,217 km) is also one of the bloodiest borders in the world. India's primary border guarding organization, the Border Security Force (BSF), is "notoriously violent," known for its systematic attacks not only on border crossers but also on the local people on the Indian side. Torture, extrajudicial killings and forced abduction are regular features at the border. These X-ray photographs are a testimony to state violence against people in the borderlands. The incidents occurred from June 2015 to February 2023. The names have been changed to shield identities.

All photos by the author.
Courtesy of Partha Sengupta.

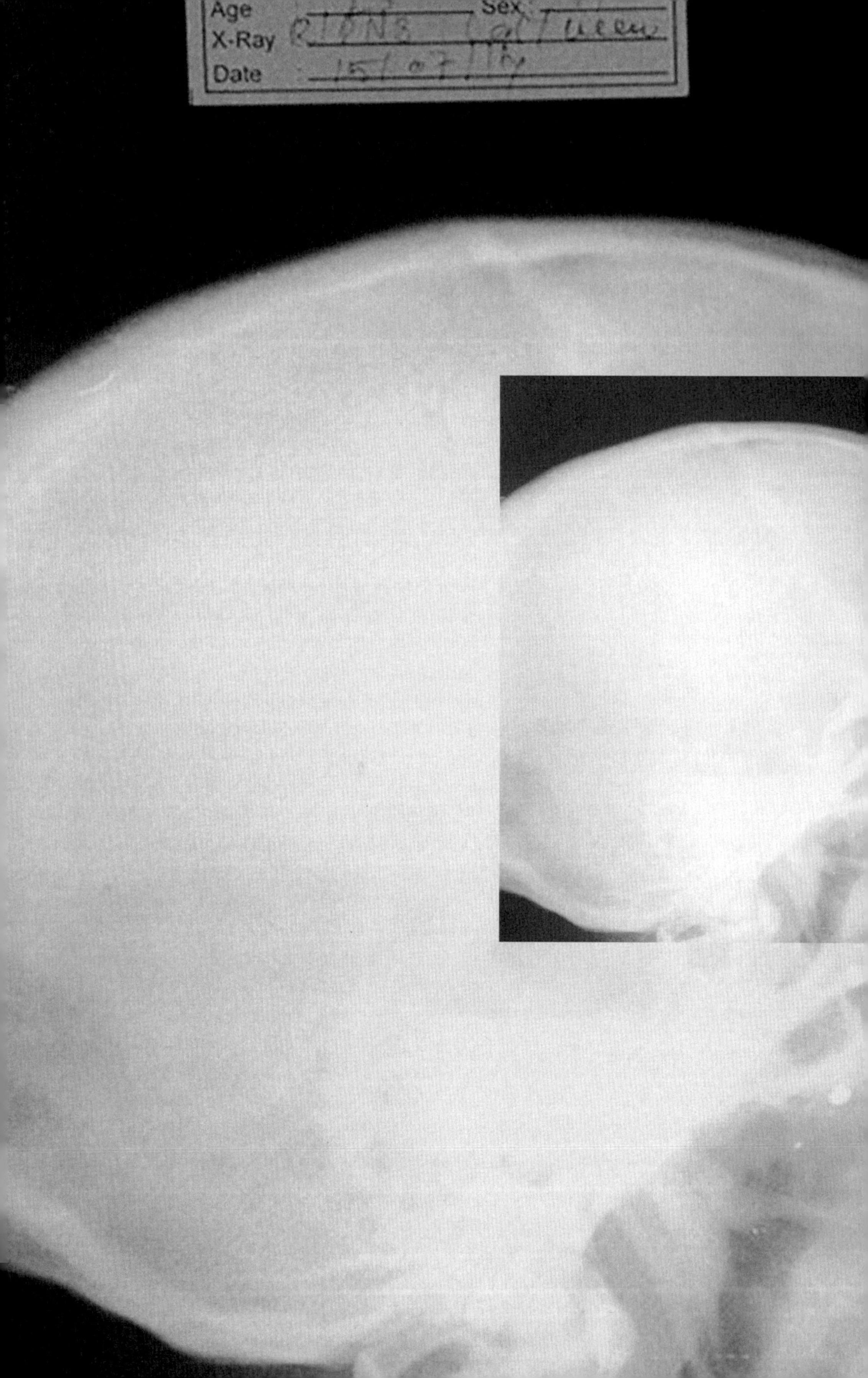
Age
Sex
X-Ray
Date

Fig. 2 — Hossain Ali is a visually impaired person because of a shotgun injury that damaged his eye, fired by an Indian border guard.

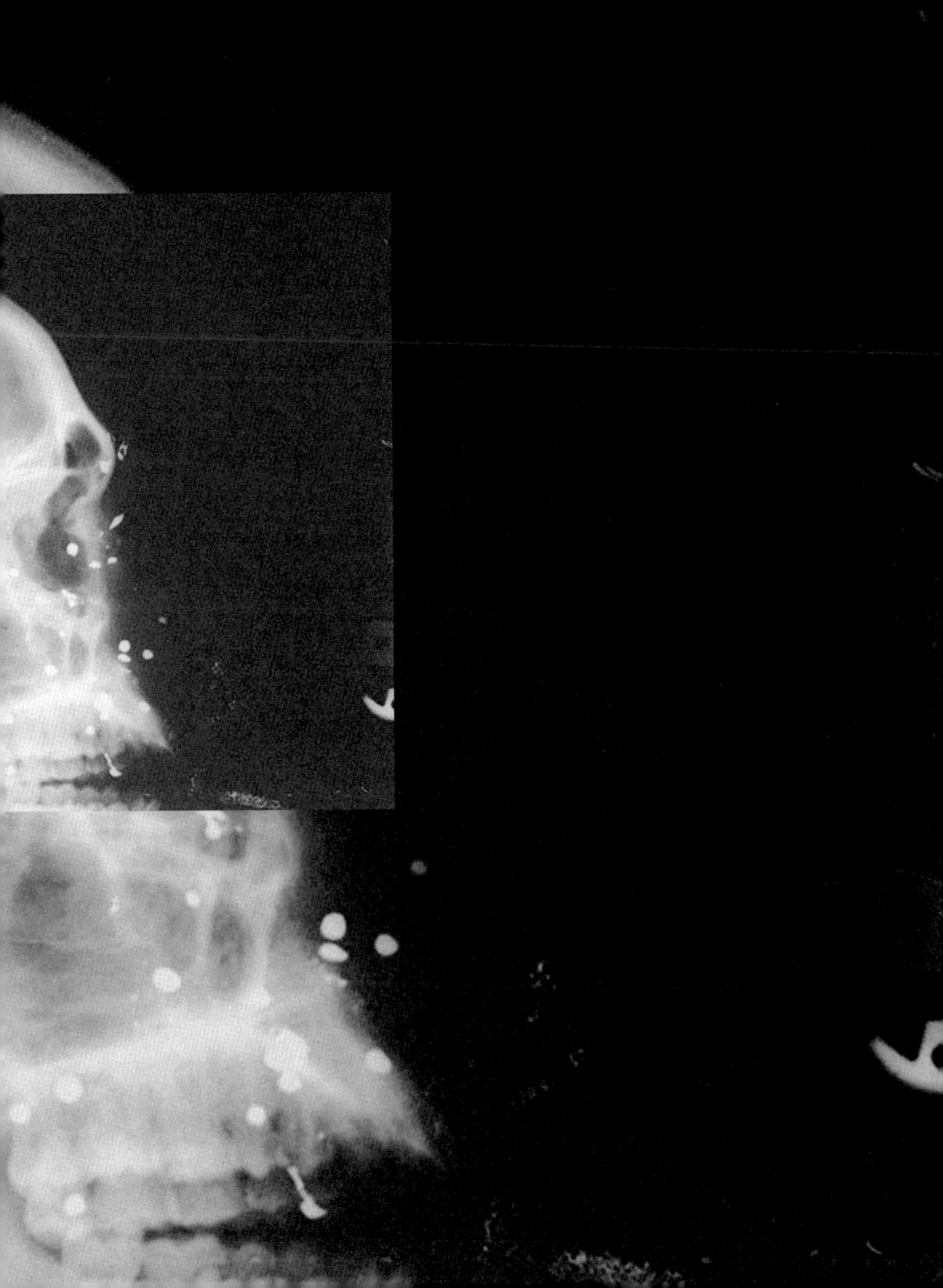

Fig. 3 — The white dots are shotgun pellets, seen on the X-ray plate of Afzal Biswas. An Indian border guard shot him at close range, which resulted in the amputation of his right hand.

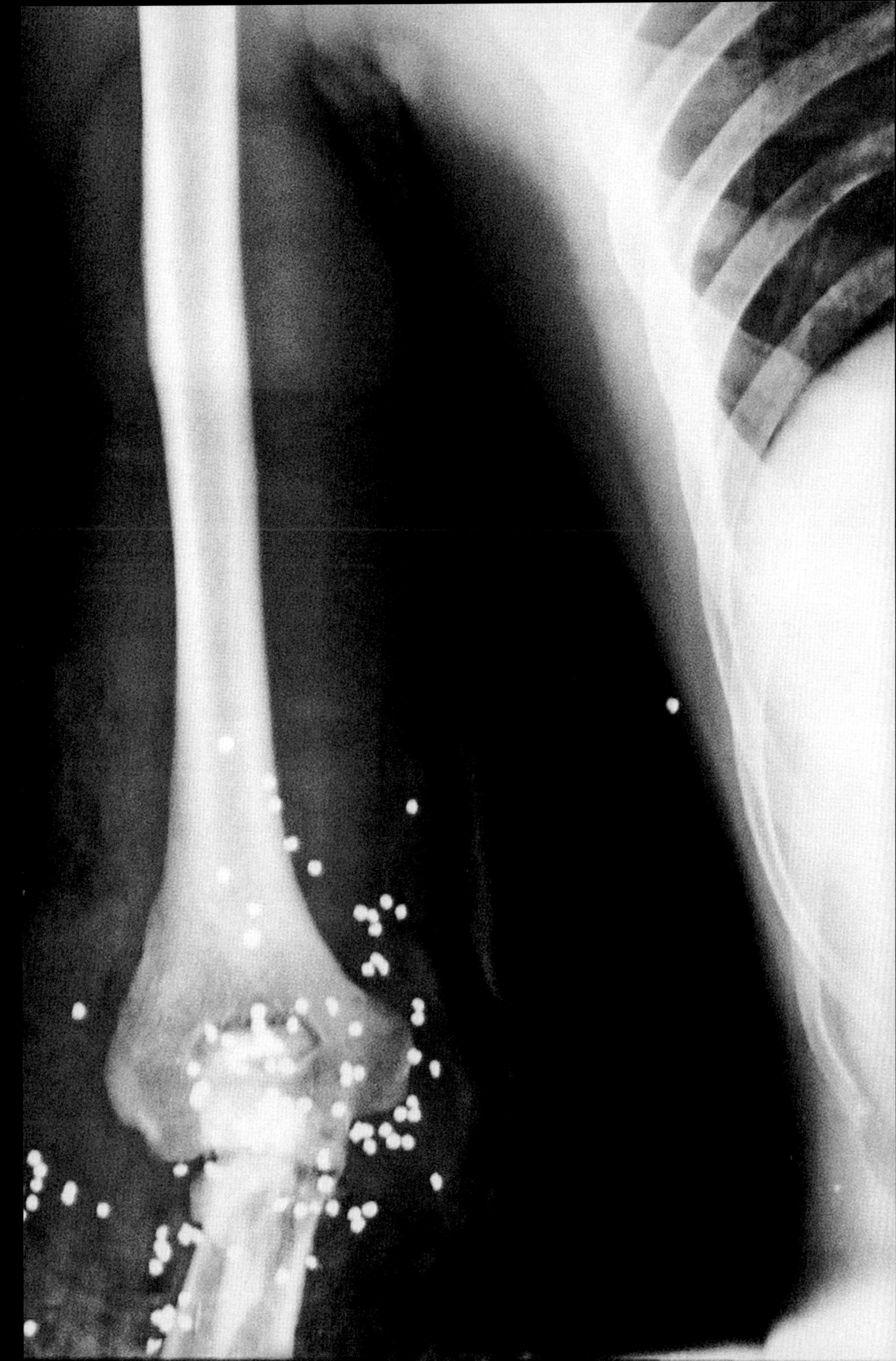

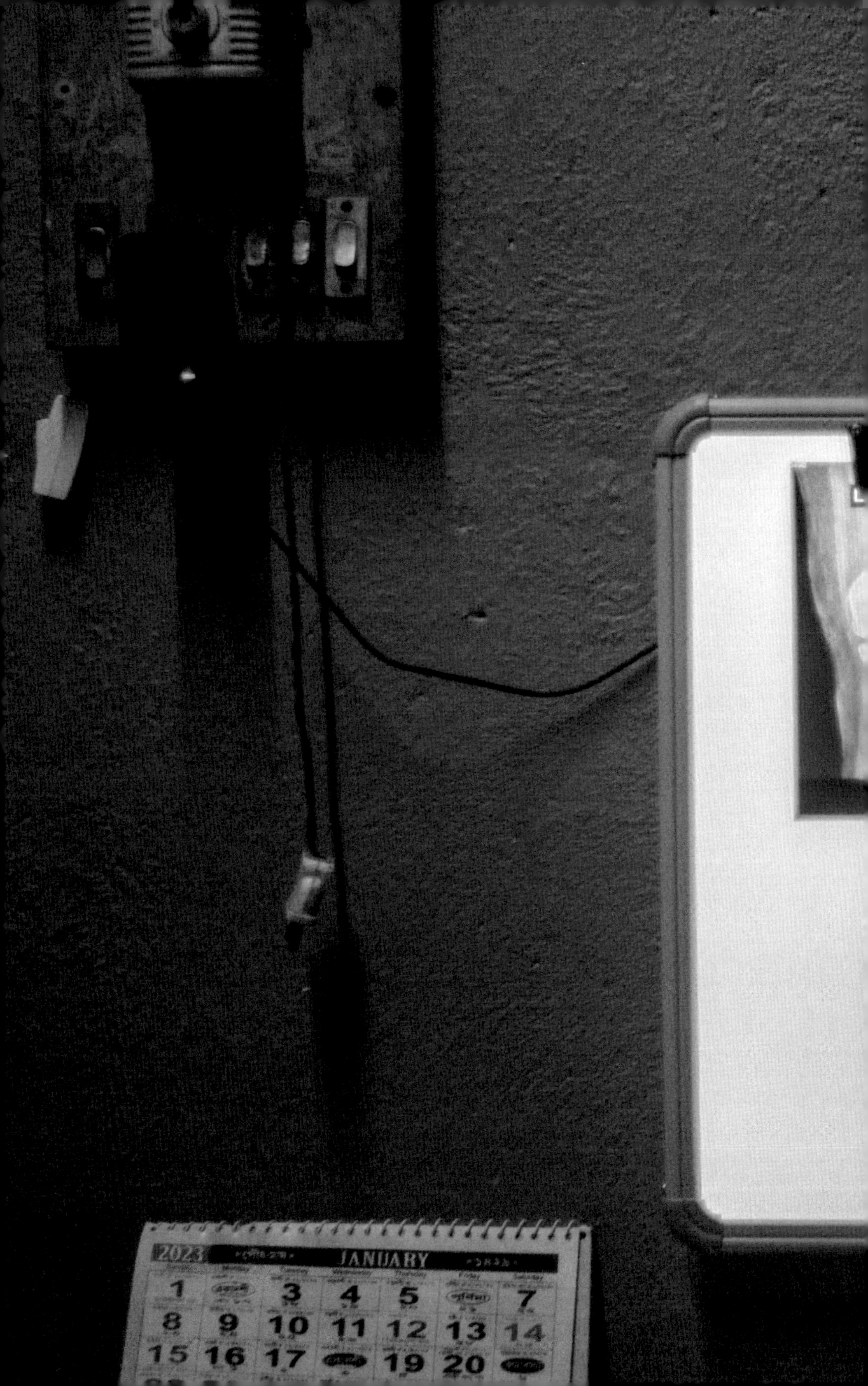
2023
JANUARY

Fig. 4 — The X-ray plate of Taslima Khatun is on display in a doctor's practice, showing a broken knee caused by the beating of an Indian border guard.

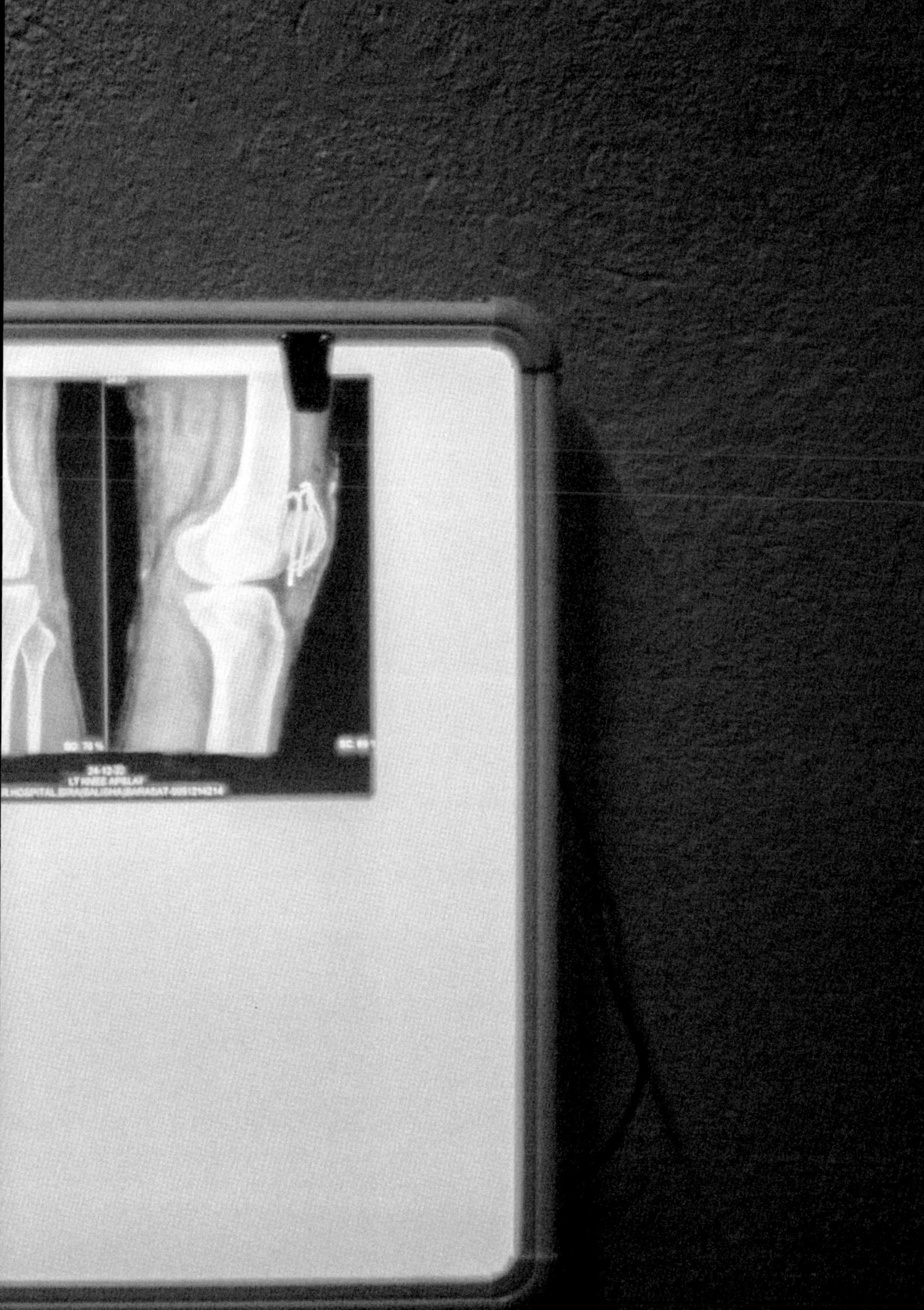

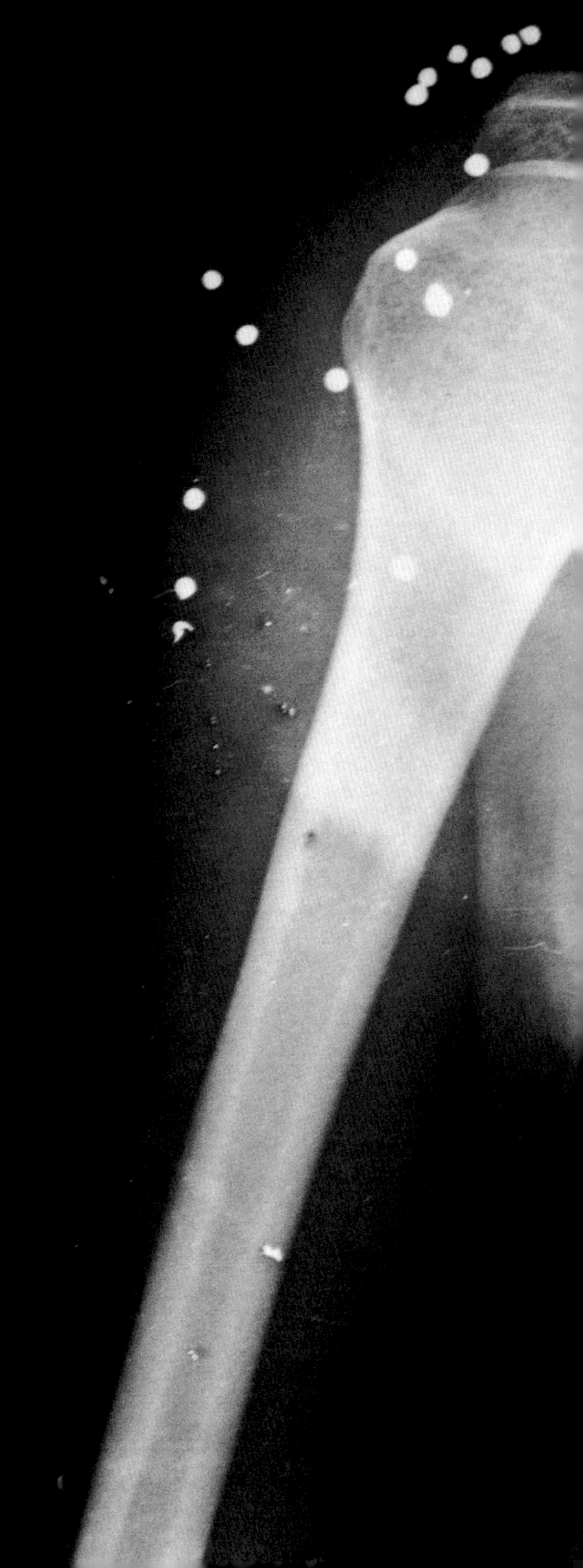

Fig. 5 — The white dots are shotgun pellets, seen on the X-ray plate of Hossain Ali. An Indian border guard shot him at close range.

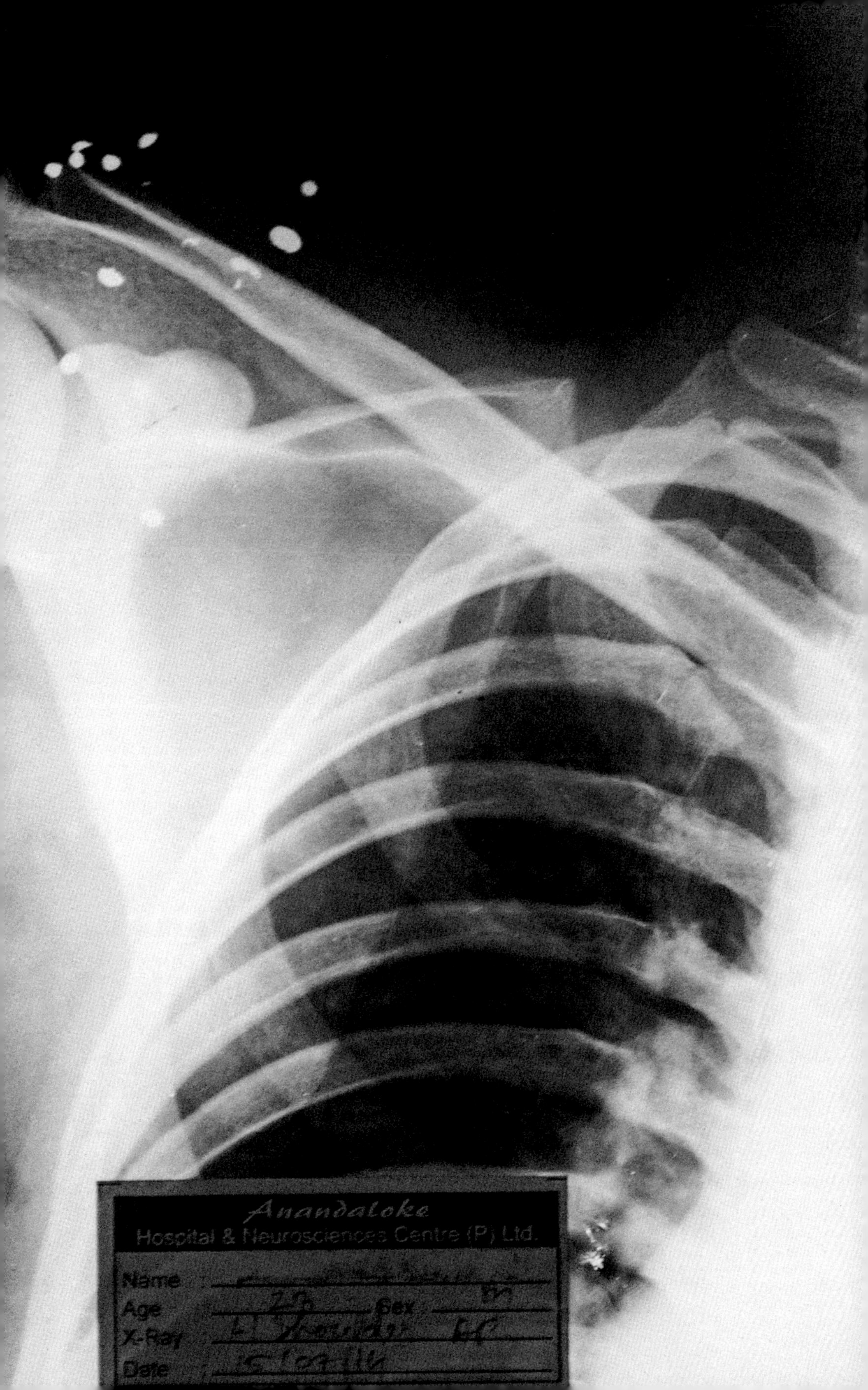
Anandaloke
Hospital & Neurosciences Centre (P) Ltd.
Name
Age
Sex
X-Ray
Date

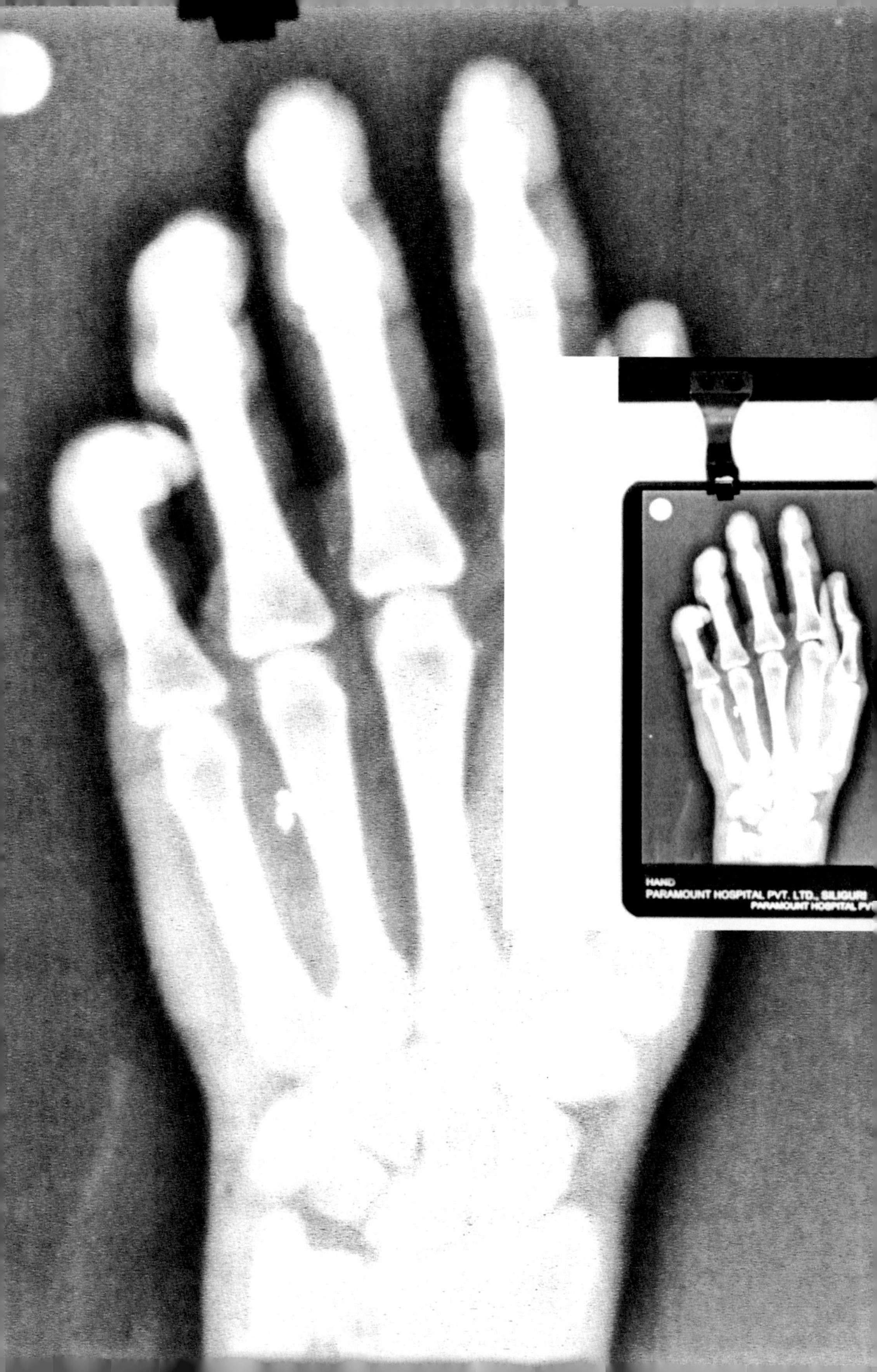
HAND
PARAMOUNT HOSPITAL PVT. LTD., SILIGURI
PARAMOUNT HOSPITAL PVT

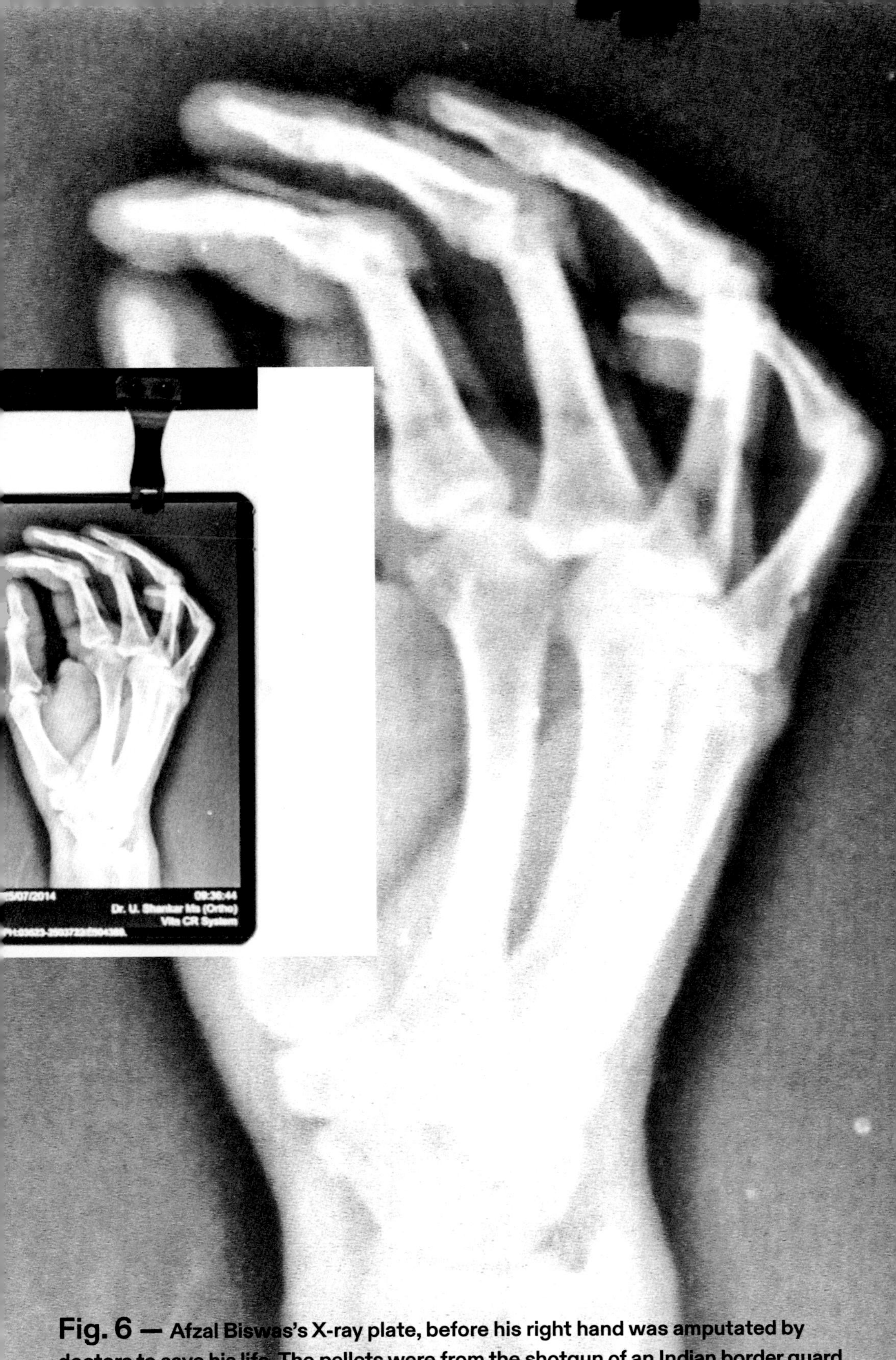

Fig. 6 — Afzal Biswas's X-ray plate, before his right hand was amputated by doctors to save his life. The pellets were from the shotgun of an Indian border guard.

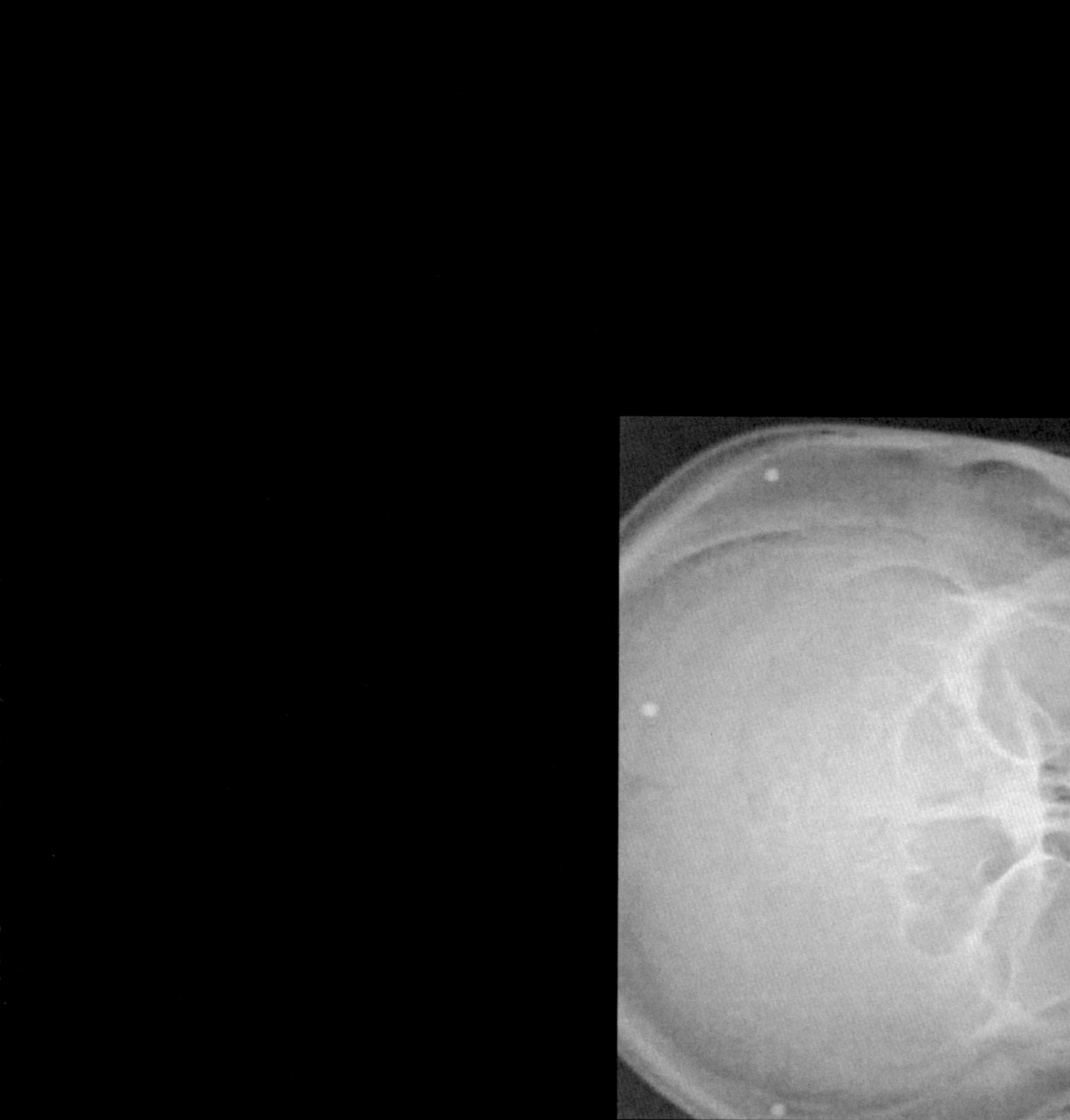

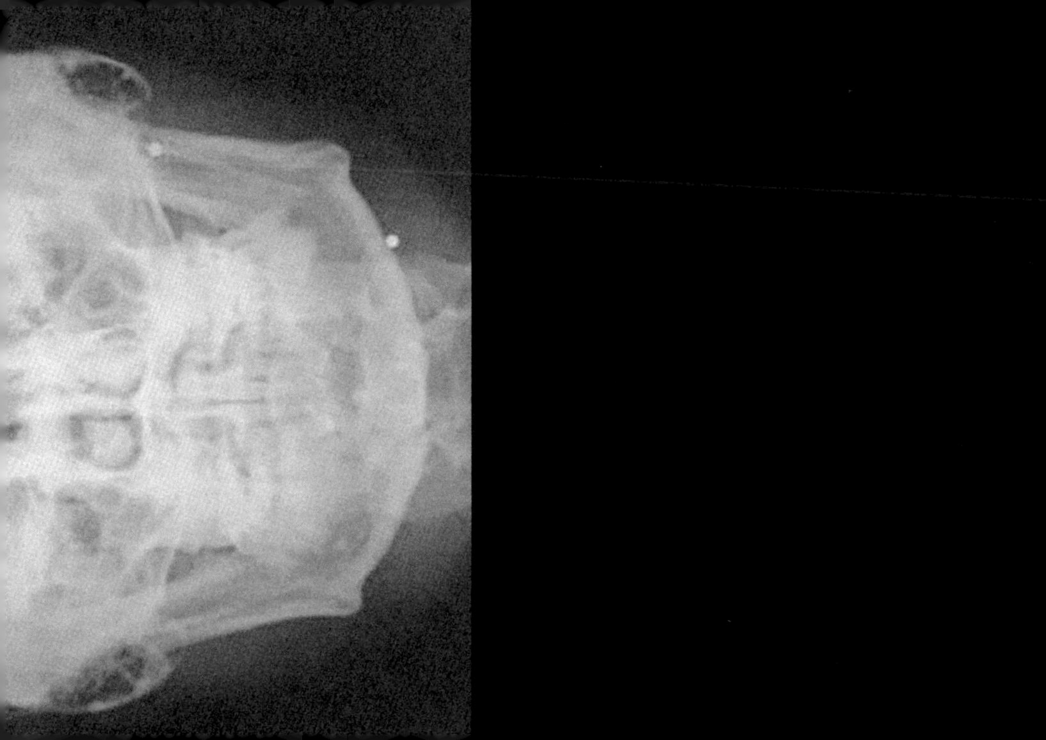

Behzad Khosravi Noori

In the Dermis of the Body

This essay delves into the premise of distance, encompassing physical separation from home and distance from physical violence but attached to the representation and imagery around it.

I see myself as an observer from a distant land, displaced from the tangible experiences that shape my understanding. It is an experience that makes me realise once again that I am in exile. My only lifeline to the connection to home is the images that transport me to faraway lands, the places I am from—places I call home.

I have made a collage from a collection of various images that have circulated across the internet during the recent *Zan*, *Zendegi*, *Azadi* women's uprisings—an embodiment of the collective struggle for freedom and empowerment. The dark blue hue adorns a plastic backing, concealing a layer of silver halide crystals beneath it—evidence of the X-rays that are silent witnesses to pervasive state violence. The amalgamation of images in the collage, united as one body, is a testimony to never-ending violence; a site for melancholic reflection on the transience of human and material existence.

The collage and this essay together are a speculative itinerary attempting to bridge 'here' and 'there,' an attempt to seize hold of memories before they are lost to the abyss of oblivion. It is the testimony of the distance from the surface of the body, where radiation transcends the skin and delves into the depths of the dermis, where the concealed traces of violence find their voice testifying to the hidden, the invisible and the distance; the fragment of the image and imagination. It serves as fragile threads that attempt to weave together the emotional and physical gaps. The dissemination of X-ray imagery in Iran coincided with a series of nationwide civil protests there, known as *Bloody November* or *Bloody Aban* (in Persian: Aban-e khonin), which occurred during 2019 and 2020. Subsequently, in 2022, a surge of additional images emerged following the tragic death of Jina (Mahsa) Amini while she was being held in police custody in Tehran. Her untimely death became the catalyst that fused together decades of seething resentment and suppression endured by Iranian women.

The Shooter in the Black Jacket

I immersed myself in the realm of short video clips shakily recorded by a mobile phone, some captured from the top of balconies or half-open windows—in the distant realm of images, where revolution unfolds. I am bombarded with images. My smartphone has been transformed into a gateway for aimless scrolling. Delving into the images and short clips, it becomes painfully clear that it offers a shallow glimpse of the intricate tapestry of experiences in those distant lands.

Men armed with firearms took aim and the unmistakable sound of gunshots echoing through the air would follow. The voices on the opposing end would be a cacophony of screams of anger, fear, and hatred. Some videos depicted the shooter, occasionally adorned in the attire and headgear of law enforcement but more frequently donning the uniform associated with the paramilitary forces known as *Lebas e Shaksi*. Clad in military pants, a shirt draped over them and sometimes black leather jackets. The shooter would glance back toward their targeted prey, their victim for the day, the victims of these days.

The shooter is a young man. The shooter doesn't target a solitary victim; they employ a new technique, showering bullets upon multiple individuals' heads, using

Fig. 1 — Some bullet kills, some that come and do not remain fixed in the body and move. Those who are thrown out by the body. And those which become part of the body resting on the dermis of body untraceable, the hidden, the invisible.

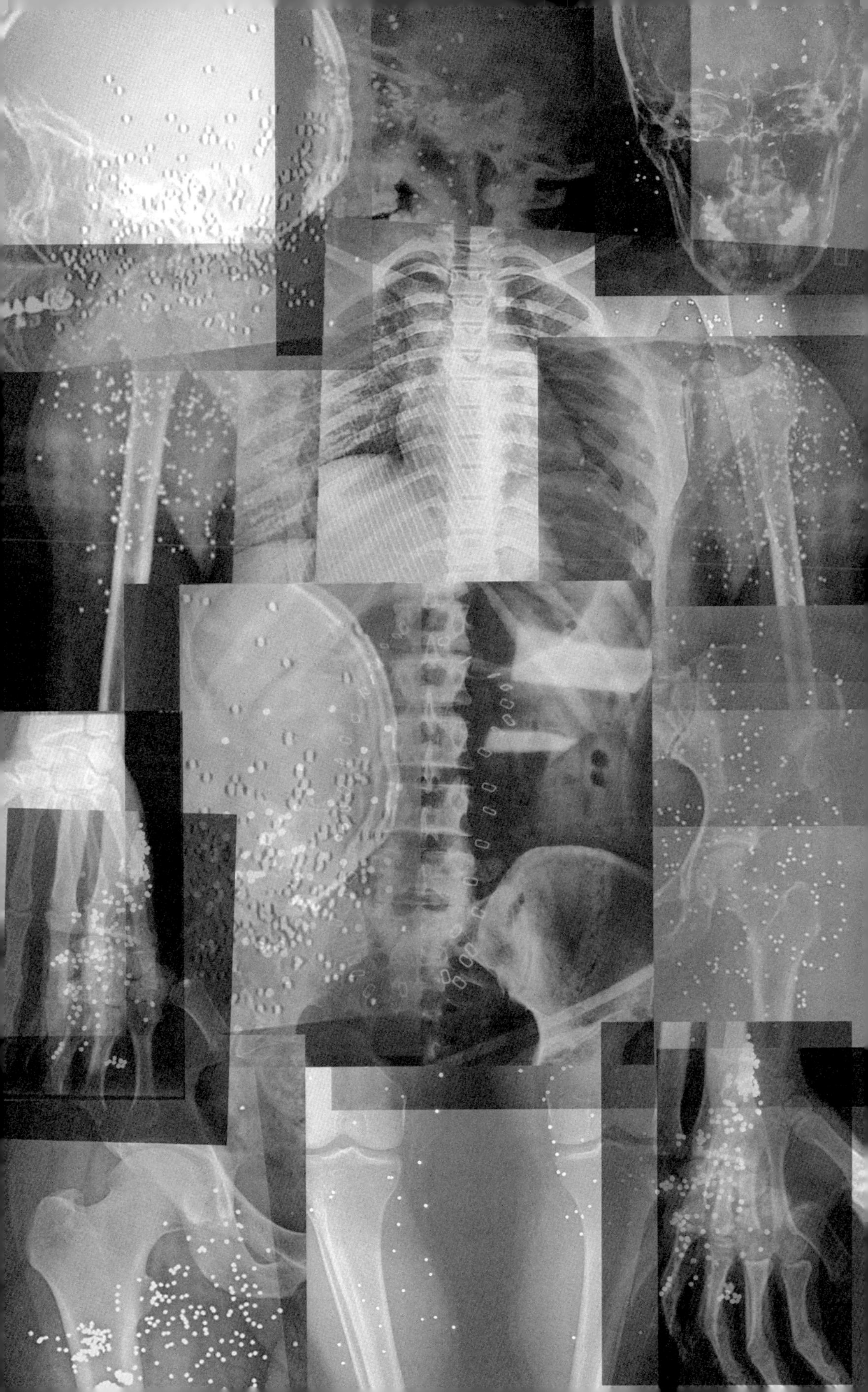

their shotguns to issue slugs carelessly. After shooting, he does not advance towards his goal; after a fleeting moment of observation, ensuring the outcome of his action, he would pivot away and gradually depart from his target with measured steps. Steps that emphasise the need for the shooter to maintain a certain distance for fear of losing control over the narrative and the power dynamics at play.
Stepping back, his gaze lowers and he immerses himself in the gun's mechanics, or perhaps he feigns indifference as if nothing of significance has occurred.
On the other side, unseen in the frame, voices of anger, hatred and longing reverberate from the surrounding vicinity. The gunshot silences the shouts of *Zan, Zendegi, Azadi.*
I can almost imagine the acrid scent permeating the atmosphere—a blend of tear gas and freshly spilt blood. Witnessing these images evoked a burning sensation in my throat and tears welled up involuntarily in my eyes, a poignant reminder of experiences to heed.
The camera person's trembling hand signalled the video's conclusion as they retreat from the window, abruptly interrupting the clip at thirty-eight seconds—an end without an end.
In this state of hopelessness, I ask again and again, what happened to the victim? Did they survive? Are there any images? Witnesses? A sensation of pain surges inside my chest, coursing its way up towards my head. I search again for more information; I ask friends. One of the injured survivors recounts their harrowing experience,

> I rushed towards the police force. I threw back the tear gas canister they had fired at us. Suddenly, there was the sound of a big explosion but I ignored it and kept going, even though my lower body was in unbearable pain. After ten agonising minutes, my knees and thighs burned with intense pain. I rolled up my pants and saw that I was bleeding heavily. It was clear that I had been shot multiple times from the waist down, making it impossible to walk any further. A friend and some other people came to help me. They moved me to my home. Later, I understood that I had been hit by around eight pellets, each one lodged inside my body. I went to the hospital for an X-ray but the fear of being arrested made me leave the hospital. The bullets remain in my right thigh and knee, leaving me unable

to walk. I was wounded below the waist. Others were shot in the face and chest. The militias showed no mercy, using black Winchester-style rifles loaded with pellets while the special forces launched their attack.[1]

The X-ray of Birdshot

In his 1728 Cyclopaedia, Ephraim Chambers documents the shotgun as a weapon for avian extermination.[2] Throughout the latter half of the 18th century, these bird-killing instruments found their way into the arsenals of England and the United States of America. However, after the First World War in late 1917 and the emergence of labour movements, shotguns and their ammunition became instruments of both bird hunters and law enforcement officers.[3] Is there a metaphor here? The cliché relationship between birds and protestors? Flying as an indication of emancipation? Since 2010, pellet shotguns have been used by anti-riot police in Bahrain, Egypt, Tunisia and Iran to disperse protests. Similarly, in the Indian state of Jammu and Kashmir, law enforcement authorities have also used pellet shotguns since that time. They are locally referred to as pellet guns or pump action guns and are described as a supposedly "non-lethal" means to control protesters.[4]

Buckshot shotguns were imported and introduced to Iran for military purposes prior to the 1978 revolution. In 1977, the influence of the revolution seemed far-fetched, at least for the Shah. The police, initially tasked with dealing with everyday crimes, now took on the role of suppressing demonstrations. While the military boasted ample resources and equipment, the police force lacked anti-riot gear and training. To fill the gap, 870 Remington shotguns from the United States were obtained to equip the police

[1] Personal phone communication with an eye witness.

[2] Ephraim Chambers, *Cyclopædia, or, An universal dictionary of arts and sciences* (London: J. Knapton, 1728), 88.

[3] Scott W Phillips, "A Historical Examination of Police Firearms", *The Police Journal* 94, no. 2 (1 June 2021), 122–37.

[4] "Losing Sight in Kashmir: The Impact of Pellet-Firing Shotguns", *Amnesty International USA*, https://www.amnestyusa.org/reports/losing-sight-in-kashmir-the-impact-of-pellet-firing-shotguns/. Accessed 31 August 2023.

force. However, these weapons were delayed reaching the hands of the Shah's police force in time and couldn't be used during the 1978 uprising.
After the 1978 revolution, shotguns become one of the most frequently deployed weapons to suppress protests in Iran. The shotguns used by the police force are manufactured in Iran, Turkey and the United States, while the bullets are made in Italy and France.[5]
The impact of shotgun injuries remained hidden from sight until the advent of X-ray technology by W.C. Röntgen in December 1895. With his invention, Röntgen discovered that X-rays could penetrate human tissue, revealing the underlying bones and tissues, the image from the dermis of the body. The news of this discovery rapidly spread across the globe, capturing the attention of medical professionals worldwide. Within a short period of time, doctors in Europe and the United States began using X-rays to identify gunshot wounds.[6]
But what does W.C. Röntgen's invention show us in these X-ray images from the current uprising in Iran? The compelling visual evidence captured and circulated reveals the extensive and alarming use of live ammunition fired from shotguns against protesters in Iran.
The vast majority of the images present the concentration of the pellets in one specific part of the body. Numerous photographs depict police and militia using these weapons at distances of less than five metres, aiming specifically for the heads or chests of protesters. Bullets discharged from close range, mainly targeting vital areas such as the head, chest or pelvis, can be fatal. The prevailing consensus emphasises the necessity of shooting from a distance rather than at close range to avoid killing people. However, in these instances, it appears that pellet bullets were being used as lethal ammunition.
X-ray images of injuries from recent years in Iran serve as further evidence of state violence.

[5] Omid Shams, "Special Report: What Equipment Is Used To Suppress Iran Protests, Which companies Provides Them?", *Iran Wire*, https://iranwire.com/en/politics/109507-from-armored-vehicles-to-kalashnikovs-equipment-used-to-suppress-iran-protests/. Accessed 31 August 2023.
[6] Cf. Alexi Assmus, "Early History of X Rays", *Beam Line, A Periodical of Particle Physics Summer* no. 2 (1995), 10–24.

An Unwanted Guest, an Indissoluble Connection

The injured protestors are frightened to seek hospital treatment for fear of being arrested. Consequently, the majority of X-ray images were taken in private clinics, where there is a lower risk of detection by the authorities. Such is the extent of their terror that some victims, gripped by fear of identification and subsequent arrest, choose to endure their pain alone, refusing the solace of medical treatment and surgical intervention. The pellets embedded in their bodies are unwelcome reminders of the violence they endured, lying beneath the surface of the skin in the dermis of their bodies.

The bodies bear the weight of lodged pellets, each piercing deep within their flesh. Trapped within the confines of these wounds, the bullets become unwelcome inhabitants, forever altering the lives of their victims. In this realm of secrecy and trepidation, the only remnants of this harrowing event are captured in the form of X-ray images.

Within the X-ray images, a horrific sight unfolds—a skeletal figure surrounded by a hollow, evidence of the merciless bombardment by white pellets, leaving no room for interpretation. These slugs, driven by malevolence, relentlessly damage the bodies they penetrate, leaving behind a trail of devastation and trauma. The juxtaposition of the faceless images amplifies these victims' anonymity and collective suffering. Stripped of their individual identities, they are reduced to mere symbols of resilience and endurance in the face of unrelenting violence. Thus, the buck shots become ominous reminders of a harrowing past and of the ongoing predicament of state violence within the confines of these visceral and terrifying records.

Some pellets manage to navigate their way out of the body. Some bullets remain in their original position, nestled within the tissues between layers of flesh, skin, and bones in the dermis of the body. These static bullets establish a peculiar presence as if they have become an inseparable part of the body's composition.

The bullets that persist within the body create a symbiosis with their host. This intricate bond forms an indissoluble connection, with the bullet becoming an integral element of the body's existence.

The body becomes the pellets' host and the pellets are part of the body. A bullet in the body is a memory of an event which leaves indelible marks etched into these bodies, forever resonating with the pain and trauma endured by the victims, a manifestation of agonistic co-existence.

Once the healing process is complete, there may be no visible evidence of the injuries inflicted. The permanence of the pellet wounds only becomes apparent when X-ray images are examined. These images demonstrate the existence of bright white slugs embedded in the dermis of the victims' bodies. Only through this diagnostic technique can the enduring presence of the pellets be observed, representing a lasting testament to the violence endured. In the X-ray images, pellets are not only visible but also tangible when they come into close proximity to the skin. Bumps beneath the skin become apparent, scattered here and there. While the skin can often heal itself, leaving only minor or even invisible scars on the victims, the eye lacks this capacity for self-healing. After witnessing the haunting X-ray images of injured dissidents, how can we possibly erase such memory from our minds? How can we continue pretending it never happened or simply turn a blind eye? The weight of that memory will forever linger, its imprint etched into the depths of our consciousness.

Some bullets kill, some enter and do not remain fixed in the body but move, and some that become part of the body resting are untraceable, the hidden, the invisible traverse beneath the skin's surface to witness the unfolding revolution, the indelible testament to violence etched into the dermis of bodies.

References

Amnesty International. "Losing Sight in Kashmir: The Impact of Pellet-Firing Shotguns", *Amnesty International USA*, https://www.amnestyusa.org/reports/losing-sight-in-kashmir-the-impact-of-pellet-firing-shotguns/. Accessed 31 August 2023.

Assmus, Alexi. "Early History of X Rays." *Beam Line, A Periodical of Particle Physics Summer* no. 2 (1995), 10-24. https://api.semanticscholar.org/CorpusID:74815630.

Chambers, Ephraim. *Cyclopædia, or, An Universal Dictionary of Arts and Sciences.* London: J. Knapton, 1728.

Phillips, Scott W. "A Historical Examination of Police Firearms", *The Police Journal* 94, no. 2 (1 June 2021), 122–37, https://doi.org/10.1177/0032258X20917433.

Shams, Omid. "Special Report: What Equipment Is Used To Suppress Iran Protests, Which companies Provides Them?" https://iranwire.com/en/politics/109507-from-armored-vehicles-to-kalashnikovs-equipment-used-to-suppress-iran-protests/. Accessed 31 August 2023.

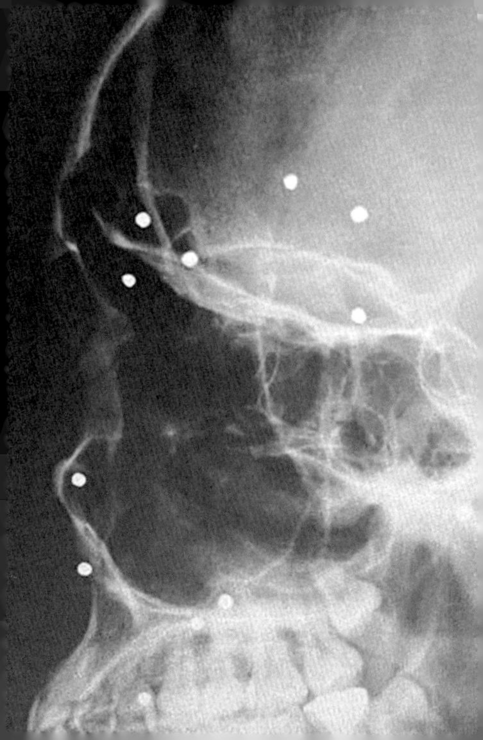

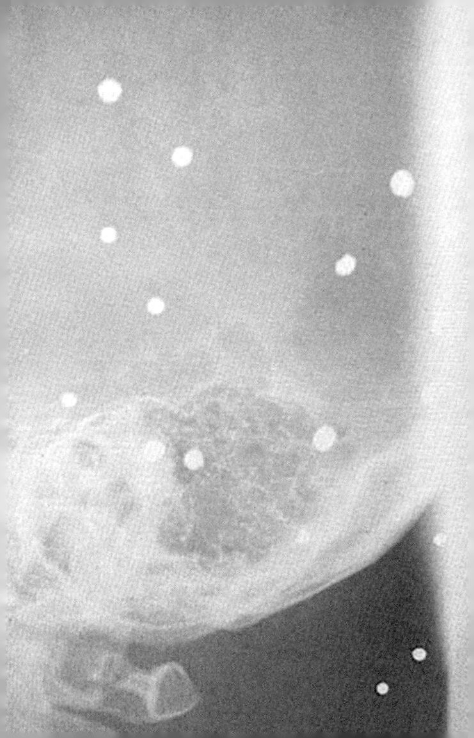

Yara Sharif

A Scarf, a Sewage Pipe and a Settler:

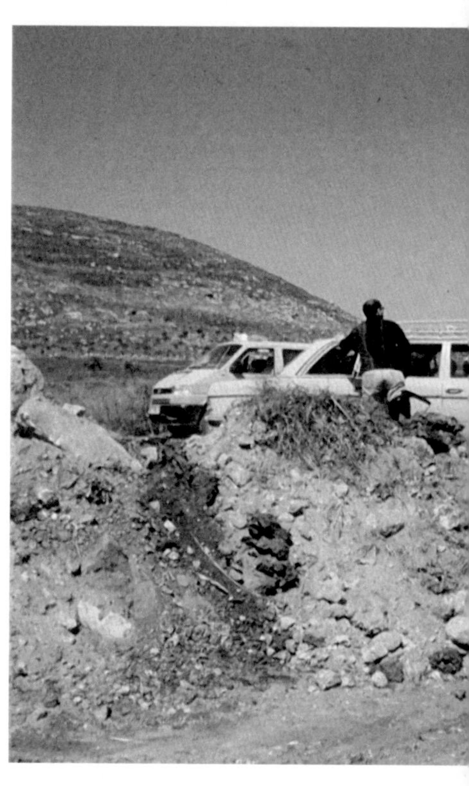

Fig. 1 — Finding an alternative route in no man's land to get to Nablus. Courtesy of Riwaq.

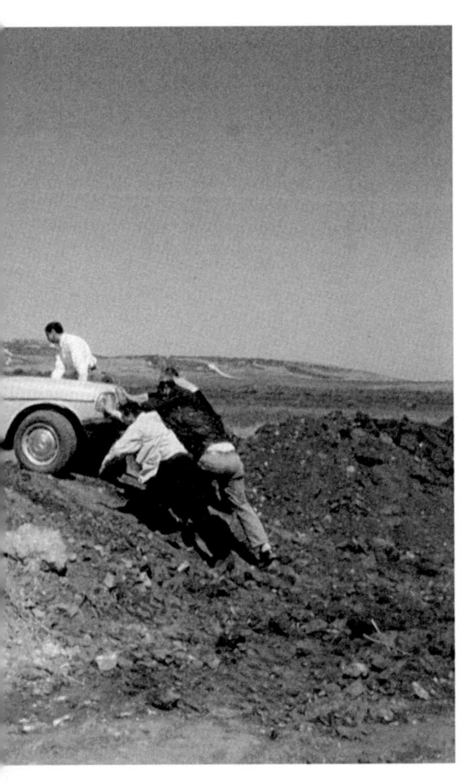

On Tactics, X-ray, and the Right to Opacity

"The orders are to let no one through today", the Israeli soldier was shouting while I was trying to drag myself and my suitcase through the dusty checkpoint. I had to find my way out to Nablus. At the edge of the road, drivers were shouting their destinations: "Tora Bora, Valley of Fire, Tom and Jerry." Mine is Tora Bora. It may sound like Afghanistan; but it's just the new informal route across no man's land to get to Nablus through the rocky landscape and dramatic dusty stone quarry.[1]

[1] Yara Sharif, *Architecture of Resistance: Cultivating Moments of Possibility within the Palestinian/Israeli Conflict* (London: Routledge, 2017), 48.

Israeli military forces intensify the control of Palestinian movement and space with endless maps, lines and artefacts of occupation and division. In this landscape that Azmi Bshara calls "the Land of Checkpoints,"[2] time and immobility have become key features in defining our everyday life as Palestinians. Waiting to cross a flying checkpoint,[3] waiting to be X-rayed or strip-searched, getting on a donkey in the middle of nowhere to find a way to get to work and crawling through sewer pipes in search for alternative routes to reach home; all reflect an absurd reality in an absurd landscape. Characterised by turnstile gates, CCTV cameras, X-ray machines and face detectors, most checkpoints are extensively equipped with devices of surveillance that strip people of their dignity and basic rights.

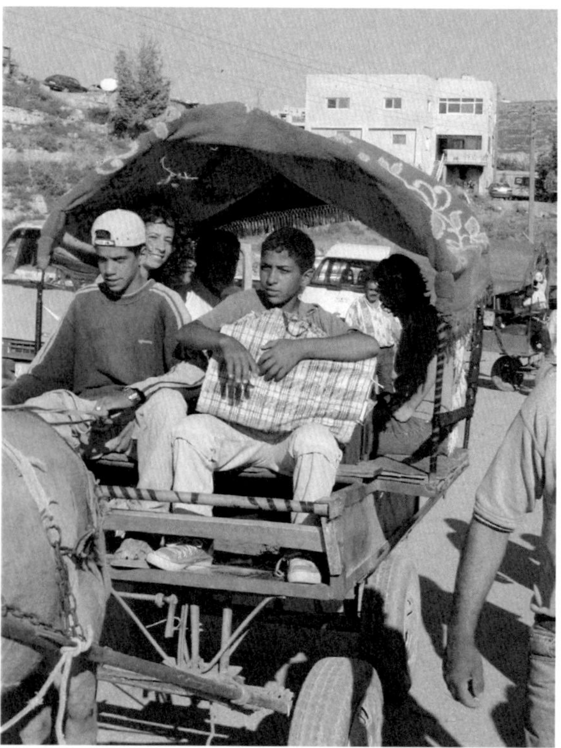

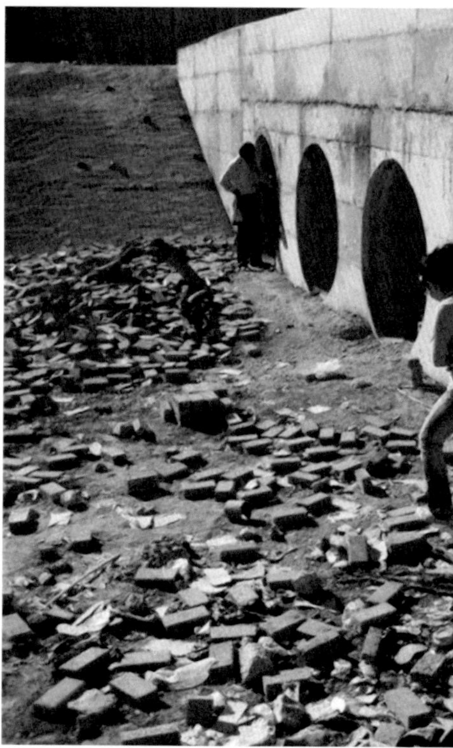

Fig. 2 — A 4-wheel, donkey-drawn cart that took us to the other side of the road. Courtesy of the author.
Fig. 3 — En route to Jerusalem through the sewer pipes. Courtesy of Dalia Hatuqa.

[2] Alexandra Rijke, and Claudio Minca, "Inside Checkpoint 300: Checkpoint Regimes as Spatial Political Technologies in the Occupied Palestinian Territories", *Antipode* 51, no. 3 (2019), 970.
[3] A flying checkpoint refers to random checkpoints erected by the Israeli occupation. They appear and disappear across Palestinian roads to disrupt Palestinians' daily routines and right of movement.

This colonial obsession with penetrative vision as a tool for control using all available devices takes place at a time in which Gazan hospitals are in chronic shortage of medical equipment and are in desperate need of X-ray machines and spare parts to provide medical care for its citizens.[4]

More than two million residents in the Gaza Strip rely on the Palestinian healthcare system. Almost half of the Palestinian population is trapped in one of the most densely populated places on Earth. Due to the siege and the economic, cultural and social blockade, Gazans have not only been denied movement outside their city since 2001, but are also unable to get their basic needs of medical care met because of long waiting times due to the shortage of equipment and the high demand. The X-ray machines are only seen from a colonial lens as tools to expose Palestinians who are constantly framed as suspects, or 'security threats' that require control.

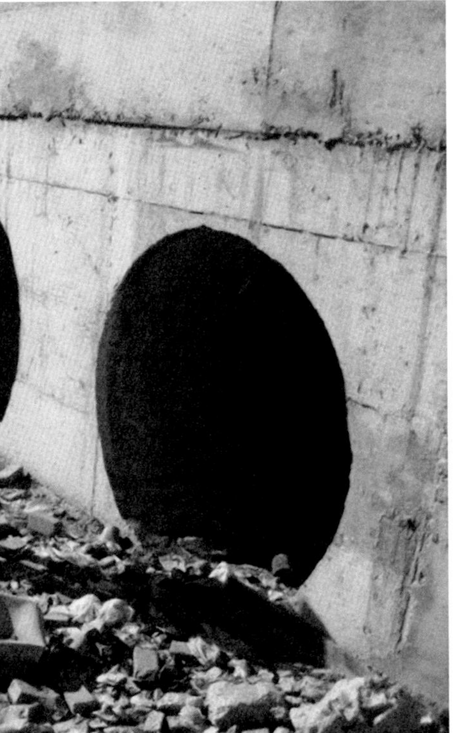

The impact of the Israeli policies of hardening the border zone not only bring fragmentation and destruction to the physical space, but also destroy the mental space and the space of imagination. Stemming from such an absurd reality, I feel obliged as a Palestinian to search for what I call spaces of possibility; ones that can stitch through my fragmented landscape and overcome the highly orchestrated matrix of Israeli occupation with its artefacts of division and surveillance.

The extract from the diary below is my tool to narrate the Palestinian landscape. While it unpacks an exhausted geography which is being contaminated with artefacts of control, it also aims to capture and narrate the sense of everyday life in Palestine with its invisible dynamics of spatial resistance that seeks through opacity to escape these devices of control.[5]

[4] Nidal Al-Mughrabi, "Gaza Says Israel Not Allowing in Enough X-Ray Machines for Medical Care", Reuters, 5 January 2023, section Middle East. https://www.reuters.com/world/middle-east/gaza-says-israel-not-allowing-enough-x-ray-machines-medical-care-2023-01-05/. Accessed 10 October 2023.

[5] The extracts are my reading of Palestine from a local lens. An alternative to the hegemonic colonial narrative that tends to deny the possibility of multiple realities. Whenever Palestinians are narrated, they are often rendered as passive subjects, in the sense that Israel makes space, while Palestinians react to it.

The Invisible Hitchhiker

(From my own diary, June 25, 2009)

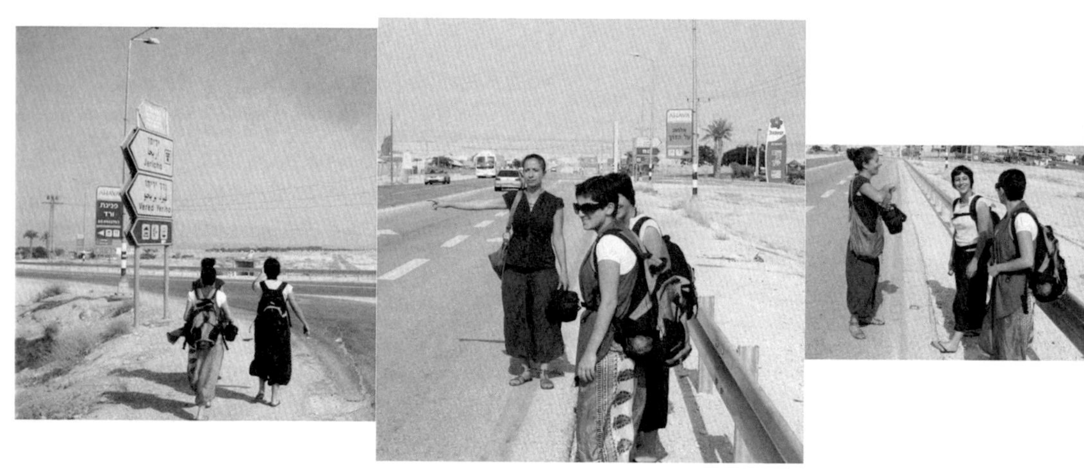

Fig.4 — The Invisible Hitchhikers waiting for a ride to the Dead Sea. Courtesy of the author.

All I wanted was to go for a swim in the Dead Sea before I made my way back to London the next day. Sahar, Dana and Luke joined in; all of them were exhausted from a hot summer day and could not wait to be in the water. We got into the famous yellow Ford van on the Ramallah-Jericho route and halfway along the van drivers in the opposite direction started flashing their lights. Apparently, a 'flying checkpoint' was blocking the entrance, just a few metres before reaching the seaside. Our driver was not interested in taking a new route, nor did he want to risk crossing the checkpoint. So we were instantly dumped in the usual 'non-place' at the junction where the roads for the Palestinians face those of the Israelis; hopefully another more adventurous driver would pick us up. From there, at the entrance to Jericho, our journey really started.

"Put your scarves on," Dana whispered while leading the way towards the opposite side of the road. "What headscarf?" I asked. With confidence, she wrapped hers around, just like the Israeli settlers do, and headed towards the hitchhiking spots known in Hebrew as "trempiyada" (טרמפיאדה). My heart started sinking; for the first time in my life, I was heading towards Israeli settlers instead of running away from them. "Dana, this is insane," I cried. "Don't worry, Yara, not only it is free, no checkpoint will stop us; we can go all the way to Eilat beach if you like." Despite my shock and fear, I sensed from Dana's confidence a lifestyle she was so accustomed to. I defied my dear friend Suad's words echoing in my heart: "there's nothing to lose but your life"—and thought to myself: "why not, let's give it a try."

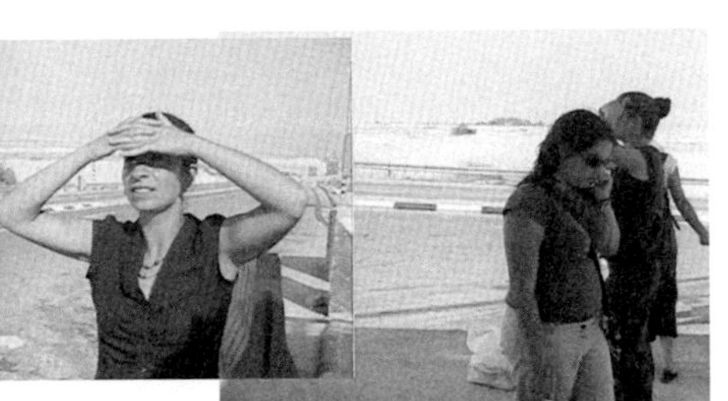

I remained quiet, trying to think about what needs to be done to pass as an Israeli settler. Apart from myself, no one else seemed to care if we got caught. Why should they? Sahar has a Jerusalem-ID, Luke is American, and Dana is so blonde that she can hardly be recognized as Palestinian—especially with the distorted stereotyped image settlers have of Palestinians' looks and images. The main problem is really me in terms of how I look. I disguised my look as much as I could. Unsuccessfully I tried to wrap the scarf around my head, while watching the crowd pointing at the cars.

Ten minutes of waiting and nothing still is coming through. Only the spoken Hebrew of the settlers around me echoing in my ears. It took me a while to feel anywhere near confident; instead, I kept examining myself to make sure I looked convincing. I knew that hitchhiking is common in Israel, but how on earth should I, as a Palestinian, know how it works? "Just press the button", Dana suggested after a long wait. Apparently, it is a way to alert Israeli drivers while waiting at traffic lights that they get ready to pick up their fellow settlers on the way. "Don't worry, Yara, we'll be fast; no settler would want to be at the entrance of Jericho at this time of the day," Dana whispered with a cheeky smile.

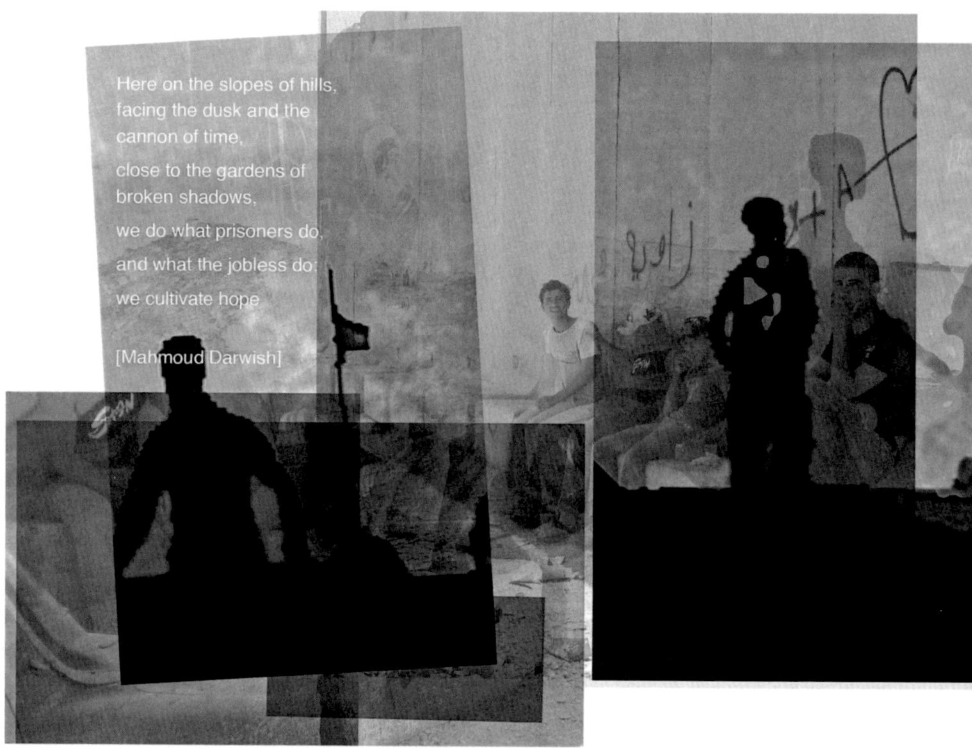

Night hunters waiting for the right moment to cross to the other side of the wall.

A few minutes later, we were all squeezed into the back of a car driving across the highway. Dana starts with her American accent, "Can we get as close as possible to the Dead Sea, please." The car driver replied, "Dead Sea? Sure, it's on my way, I live in Mitzpe Shalom. Where from are you?" While Luke tried to join the conversation, Sahar and I stared at one another realizing what we had gotten ourselves into. Mitzpe Shalom is well known for its right-wing settlers who would not hesitate to do anything if they recognized our real identity. Luke's

loud voice interrupted my scary thoughts, "We're from Ohio, but we're here to look around."
Yes, it was true—I really did start to look around, nervously examining the car, its objects and any possible signs that might help us in case we got kidnapped, attacked or had to run away. Of the whole car, I could only spot the tiny stickers on the rear window. All I could read with my broken Hebrew was, "They are handing over the Dead Sea too" and "Hebron is ours ... ever since then and forever."

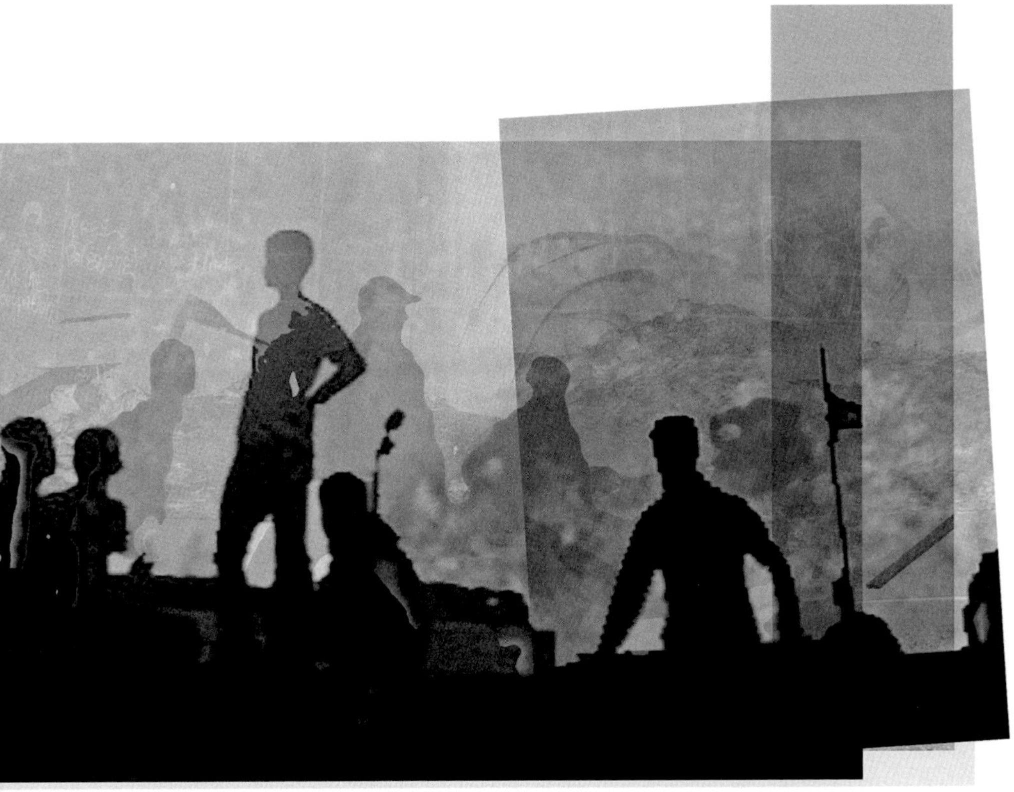

Fig.5 — Stills from *Chic Point: Fashion for Israeli Check point,* by Sharif Waked (2003). The video is an ironic reflection on politics, power, aesthetics, the body, humiliation and surveillance. Courtesy of Sharif Waked.

I happened to remember the Hebron sticker very well; Ruba my friend would use them in her car to pass through checkpoints with less hassle. Apparently, such cynical tricks have worked so well that Palestinian taxi drivers have also been using them inside Jerusalem for some time. With such stickers, not only do they get through checkpoints fast, but they also increase the chances of taxi drivers picking up customers who might otherwise not want to get into a Palestinian cab.

That map beneath the sticker also looked very familiar to me, I had once worked on the regeneration of the historic centre of Hebron for a whole year and visited every single house and walked through its maze of alleyways. I was hosted at Maha's house for three days under curfew just so that the Israeli settlers could walk free. Who knows, maybe our driver was one of them?
At that point, I felt the poster was really there to remind me that Hebron is indeed OURS—not his—just as the Dead See is also ours. If it takes an Israeli settler to drive me through my own map, then so be it; why not to enjoy it, as if eventually I would be able to cross all boundaries?
With a few jokes here and there, we were soon about to reach our destination. A shiver ran down my spine; this one however was of happiness, not of fear. For some reason, I felt that I was no longer chained; neither the driver, nor the settlements around, could occupy us anymore. At that moment, it was me who was occupying him. Just before he drove off, I asked: "Where is the best spot to get dropped at Beersheva?[6] My friend Dana wants to visit Adamama Farm."
I didn't join the final leg of the trip to Beer Alsabe', but definitely it was worth the effort. From Ramallah, Dana told me she had made it through to the village of Beit Sahour through the sewage pipe, down to Hebron, reaching her first hitchhiking point near to the Israeli settlement of Kiryat Arba (known to host one of the most extremist right-wing groups in Israel). From there she made her way to Beer Alsabe' and finally to Adamaa Farm. "I smelled the scents of Gaza and the Mediterranean few minutes away, before I made my way back hitchhiking through Jaffa, Jerusalem and finally Kalandia checkpoint." It was Dana's only way to explore what she has been missing in her whole 30 years of life.
Indeed, if it weren't for Dana's sense of humour, or the car stickers and the whole hitchhiking adventure, my own mental map would not have been stretched so far. The journey itself did not really matter; more important were the boundaries we broke with our new hidden rules of daily co-existence.[7]

[6] Beersheva is the Hebrew name of the occupied Palestinian city Beer Alsabe'. Located at the south of Palestine, it is known for its fertile soil and agriculture. Beer Alsabe' shares border with Gaza.
[7] Sharif, *Architecture of Resistance*, 64–65.

The diary is a reminder that despite the orchestrated complex matrix of Israeli military occupation, there is a degree of irony and power that lies within Palestinians that can subvert spaces of oppression into spaces of play and creativity. Despite the colonial obsession with scanning through Palestinians, their artefact of occupation with the X-ray system, cameras, checkpoints and even the physical *apartheid wall,* have resulted in trapping them in a physical and mental cage, not being able to see through what happens on the other side. With our invisible daily dynamics, it is us—the Palestinians—who occupy them. The colonial demand for transparency fails against the soft tactics of resilience and the subversive power of invisibility. The latter has in time created a public sphere in which Palestinians negotiate their space, geography and identity.

While I share these moments of 'mental liberation,' I equally collide with uneasy questions and problematic thoughts exposing my relationship with my land and identity. When there is almost nowhere left to hide, how long is this invisibility going to last?

It may be the time to return to the most basic techniques of manoeuvring our space and daily life in order to work against this highly orchestrated system of surveillance and control, the typewriter against the digital clouds, and the blind spots and marginal landscape, against the gaze of the X-ray. Nevertheless, we need to remind ourselves of the danger of this abnormality. After all, we are born free with the right to opacity!

References

Al-Mughrabi, Nidal. "Gaza Says Israel Not Allowing in Enough X-Ray Machines for Medical Care". Reuters, 5 January 2023, section Middle East. https://www.reuters.com/world/middle-east/gaza-says-israel-not-allowing-enough-x-ray-machines-medical-care-2023-01-05/. Accessed 10 October 2023.

Rijke, Alexandra, and Claudio Minca. "Inside Checkpoint 300: Checkpoint Regimes as Spatial Political Technologies in the Occupied Palestinian Territories." *Antipode* 51, no. 3 (2019), 968–988. https://doi.org/10.1111/anti.12526.

Sharif, Yara. *Architecture of Resistance: Cultivating Moments of Possibility within the Palestinian/Israeli Conflict*. London: Routledge, 2017.

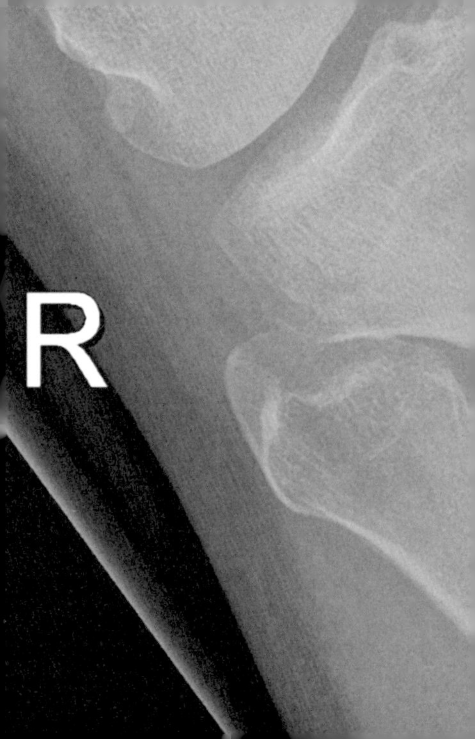
R

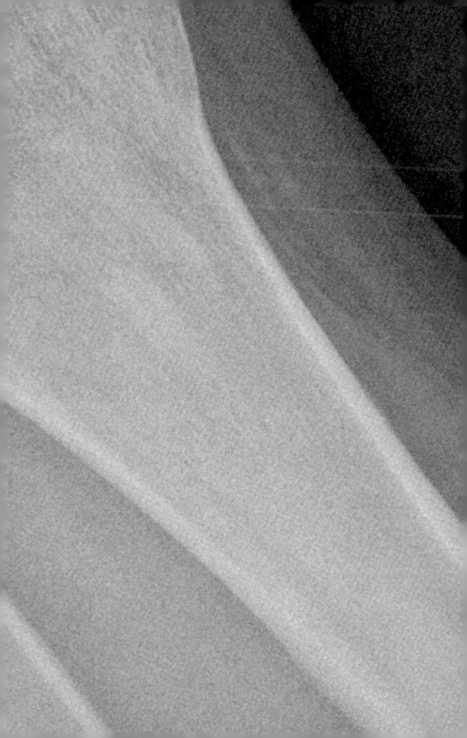

Javad Abbasi Tavallali

Silent Agony

Industrial radiography is one of the most common methods to control and to inspect checking gas pipes. In this method, X-rays and gamma rays are used to identify defects inside the pipes. Welding operations on gas transmission pipes in the South Pars gas refineries are performed with manual electrodes and non-standard tools, which means a higher risk of errors in the work and so companies use powerful devices with a beam that emits high radiation.

The workers who participate in these operations are exposed to high doses of radiation. According to experts, workers who are exposed to the ionizing rays emitted from these radiography machines are at a forty percent risk of developing certain types of cancer; the risk of nervous system tumours is fifty percent and blood cancer seventy percent.

In South Pars the workers who are in charge of the physical control of the radiography operations are called "Watchmen." According to one of the safety experts, "... everyone who is in the range of radiography operations should not be exposed to more than twenty-five microsieverts of radiation. If the amount of radiation exceeds this dose, these people are at risk of various cancers and sterility."

According to this safety expert, in order to measure the amount of radioactive rays in the body, the team should have a device called a dosimeter. Furthermore, the workers must wear radiation protection clothes. However, the private subcontractors do not provide this necessary equipment due to the rising prices that are a consequence of the US-led sanctions. The companies do not either provide needed training courses. According to a Watchman, "Only engineers and heads of the company have access to dosimeter devices and special clothes, and none of the Watchmen have had the necessary equipment."

To protect other workers, the radiography operations are conducted during night shifts. However, Watchmen and night-shift workers are not the only ones who are exposed to the dangerous rays. Drinking water containers placed on the site are also unprotected and are exposed to dangerous levels of radiation. Thus, all workers who drink water from these reservoirs are exposed to carcinogenic radioactive rays.

References

Excerpts from *Ranj Biseda* (Silent Agony), a photo report on the working conditions of labourers in the South Pars field in Iran was published online in 2020. https://www.radiozamaneh.com/500248/. Accessed 10 October 2023. The South Pars field is a natural-gas condensate field located in the Persian Gulf. Translated from Persian by Shahram Khosravi.

Javad **Abbasi Tavallali**

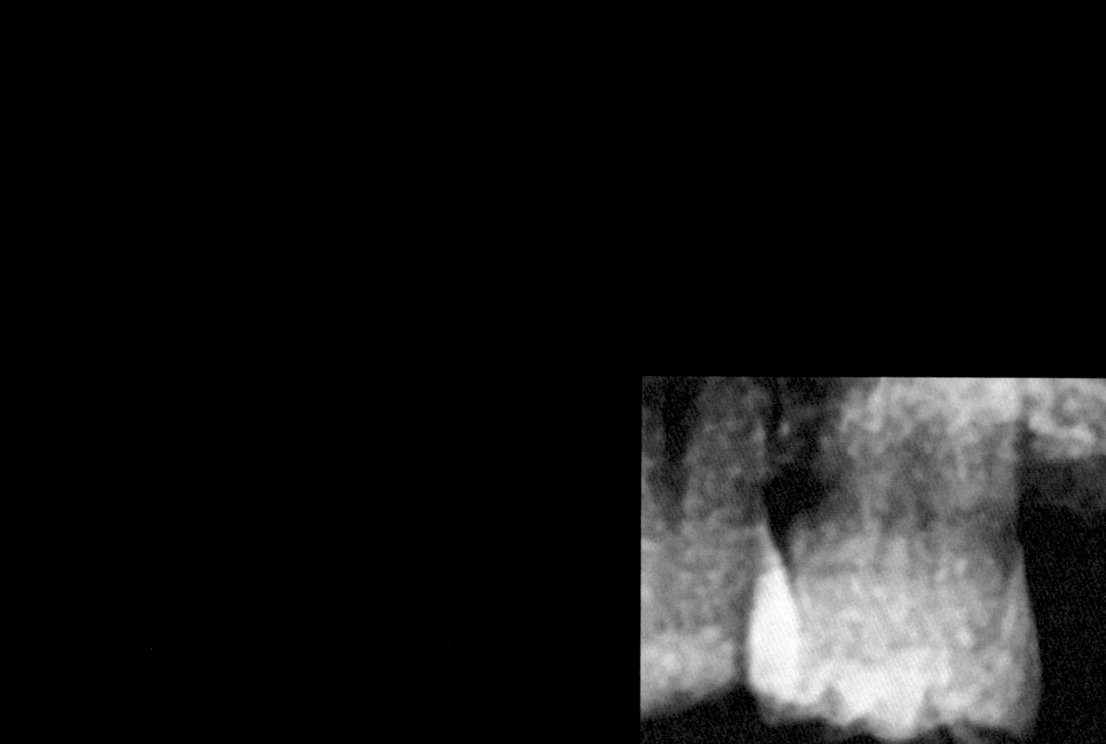

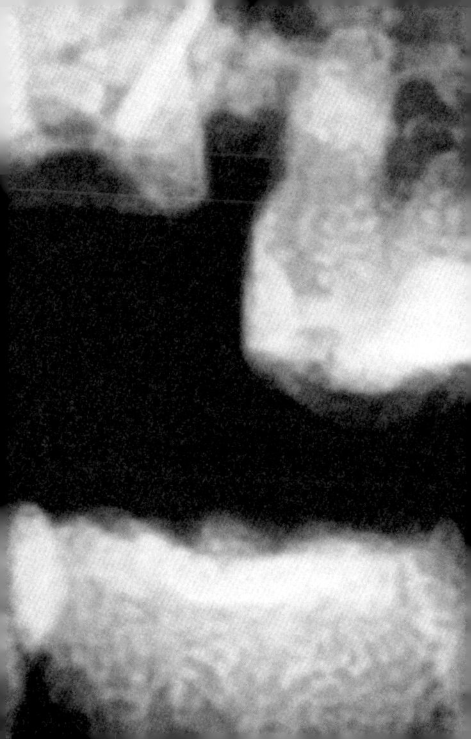

Ella Hillström and Ayodamola Tanimowo Okunseinde

A Ray in Five Folds

Prologue

In this essay, we explore the X-ray from four perspectives, which we call 'folds.' The 'folds' embody literary endeavours that increasingly venture into the realm of speculation. The first fold aims to understand the socio-political ideals that the X-ray surfaced from, to understand the imaginations and desires that formed the technological device. This foundation allows us to reimagine and speculate about alternative variations of the X-ray. The second fold probes how dance can be used to resist the X-ray and its desire to fix its subject. In fold three, we speculate about the fantasies that arose in response to the X-ray. In the last fold, we build our own speculative technology that challenges the X-ray as a techno-political object.

Fold I: Fixing

During the 19th century, physicians longed to peer into the penetralium of the human body to identify diseases but also to classify and sort the ideal body from the imperfect. Yet the body, a dense and impenetrable entity, could only be explored through hazardous methods like cutting flesh, where viscera, blood, nerves, and ligaments obscured the physician's vision, or through conducting terminal autopsies where the corpse, in its imperfection, aroused sentiments of dread and anxiety. The X-ray was therefore born out of a desire to see the living undisturbed and for 'scientists' to achieve 'objectivity', which had only begun to emerge as a scientific ideal in the mid-nineteenth century.[1] Mechanical ways of seeing, like photography and X-rays, grew hand in hand with this new ideal as they were considered capable of achieving unaltered depictions of reality. The 1895 photo of the hand of Röntgen's wife became a symbol for this new ideal vision where the inside of the subject was finally exposed.

Orthodontists were early adopters of X-ray vision.[2] The X-ray could see beyond the soft tissue of the face and gum, detecting roots, fangs, cavities, hidden teeth, and the structure of the jaw. It could also indicate the development of the lived face and skull through photographic comparison, which opened another spatial and temporal understanding of the skull and face. Yet in the early 1930s, the orthodontist Dr. Birdsall Holly Broadbent Sr. (1894–1977) saw unused potential in the technology. The X-ray, as he wrote in his article *A New X-Ray Technique and its Application to Orthodontia*, could contribute to the work of the physical anthropologists who had inspired orthodontia during the early 20th century with their "precise measurements" and strive towards "physical perfection."[3] He thanks his colleague T. Wingate Todd, who was remembered for his commitment to correlating craniofacial measurements with "craniofacial capacity."[4] Indeed, Broadbent was excited for the advancement of X-ray technology, for whilst the physical anthropologists' focused on categorizing and determining the 'ideal' measurements of the human and creating distinctions by either measuring the surface of the face and body or through dead skulls, the orthodontist with the help of X-rays could see straight through to the depths of the living face making the physical anthropologist's dead skulls obsolete.

Embedded in the ideology of the time, X-ray vision in dentistry furthered the race project spurred on by the physical anthropologists of the 19th and early 20th

[1] Lorraine Daston and Peter Galison, *Objectivity* (Princeton, NJ: Princeton University Press, 2021), 17.

[2] R. F. Mould, *A Century of X-Rays and Radioactivity in Medicine: With Emphasis on Photographic Records of the Early Years* (Boca Raton, FL: CRC Press, 1993), 89.

[3] B. Holly Broadbent, "A New X-Ray Technique and its Application to Orthodontia", *The Angle Orthodontist* 1, no. 2 (1 April 1931), 2. Ibid., "The Face of the Normal Child", *The Angle Orthodontist* 7, no. 4 (1 October 1937), 183–208.

[4] Wilton Marion Krogman, "Contributions of T. Wingate Todd to Anatomy and Physical Anthropology", *American Journal of Physical Anthropology.* Vol XXV, no.2 (1939), 153.

century who, driven by "scientific ideals,"[5] aimed to classify and organize human bodies into different races. The numerical ideals were further tangential to aesthetic ideals. Johann Blumenbach, the father of physical anthropology, named Caucasians after the people who lived in the Caucasus Mountain range, for he considered them the most beautiful. Similarly, the orthodontists of the 20th century applied X-ray to their practice both to determine an aesthetic form of "physical perfection," "normal growth,"[6] and to contribute to anthropometrics and the ideology that permeated the tradition. The differences classified by the physical anthropologists were in turn connected to inherent qualities of the person; a belief strengthened by Alphonse Bertillon who applied anthropometric descriptions to criminal identification.[7]

Many of the orthodontists of the 20th century who adopted X-ray technology were subsumed in a grander ideology that went beyond their practice. However, to measure the living head accurately and comparably, points of reference needed to be identified. Broadbent's biggest contribution was, therefore, to position the living head in a universally comparable position, and in turn identify the relationship between "internal landmarks of the face and cranial base."[8] Most importantly, to position the head in a universal position, Broadbent designed a device known as the cephalometer, which has the same name as the skull measurement device used by racial biologists to separate humans through arbitrary measurements. You might recognize its shape: two curved pieces of metal attached to one point which could open and close. On the curved pieces of metal are lines to measure differences that would later determine the race of the subject. Broadbent wanted even more precise measurements and today, the cephalometer has two connotations: an instrument for measuring the head, and a device for positioning and fixing the head in preparation for an X-ray.

[5] Amos Morris-Reich, "Anthropology, Standardization and Measurement: Rudolf Martin and Anthropometric Photography", *The British Journal for the History of Science* 46, no. 3 (2013), 498.
[6] B. Holly Broadbent, "The Face of the Normal Child", *The Angle Orthodontist* 7, no4 (1 October 1937), 183.
[7] Sekula, Allan, "The Body and the Archive," *October* 39 (Winter 1986), 18
[8] B. Holly Broadbent, "A New X-Ray Technique and its Application to Orthodontia", *The Angle Orthodontist* 1, no. 2 (1 April 1931), 46.

The cephalometer alludes to an inevitable truth; to standardize, measure and in turn differentiate, the subject needs to be fixed and positioned in time and space. As such, the X-ray apparatus and its political application could not have been realized without the cephalometer. They are mutually contingent. To transform a subject with a history and subjectivity into angular metrics that correlate to their essence, the subject needed to be fixed into one position. So, to resist the ideology that wove immobility, calculability, angularity, and distinction, we decided to dance.

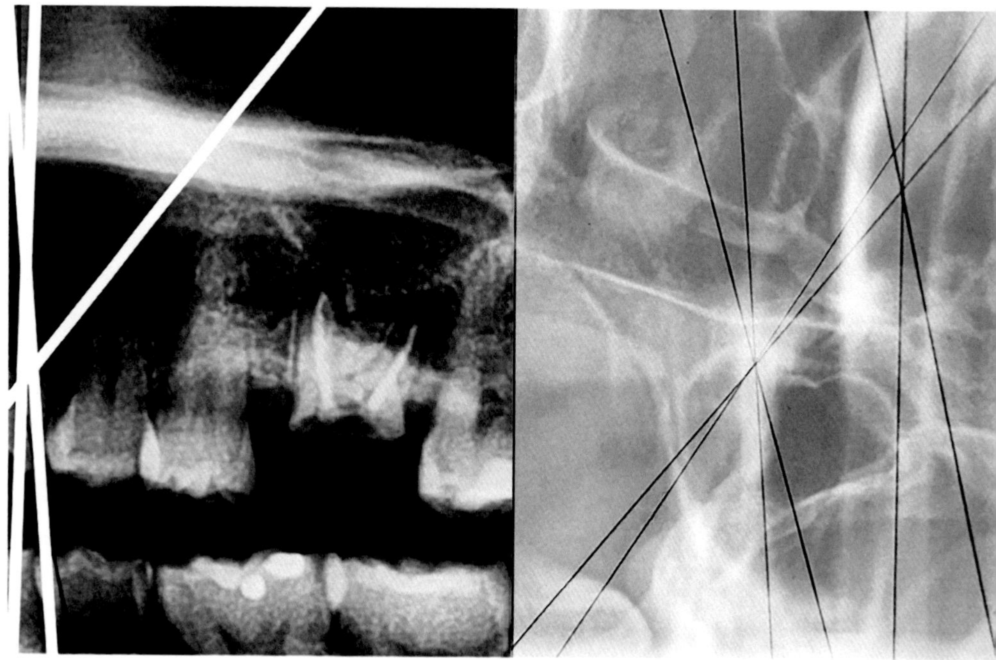

Fig. 1 – *Tatiana Overton's Molar Cyst with Frankfort Planes.* Courtesy of the authors.

Fold II: Dancing

Broadbent pushed our heads into the cephalometer. "Stay still, chin up."

Resting them in unnatural positions.

Breath, taking
shape of sound.

We
relaxed our ja
ws.

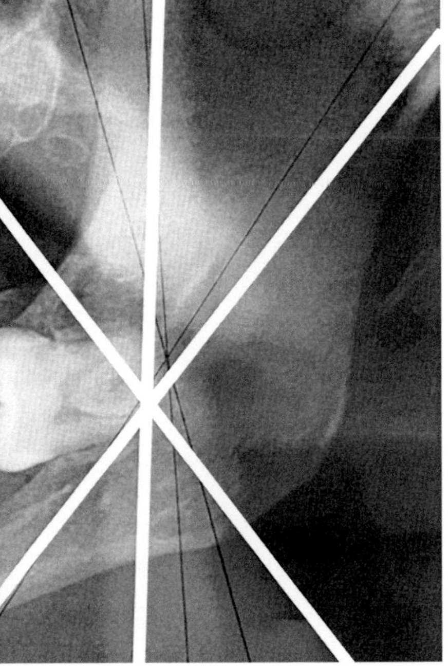

Raised our shoulders, twisted our spines.
GRIMACED: tongues
followed rhythms. Angular metrics fell in love
with fleshy uncertainty.

We shook our heads frantically; growled at the
architecture.

Fold III: Seeing

To identify craniometric points and angles that could advance the physical anthropologist's mission to classify the human, the face needed to be positioned and fixed in an 'unnatural' head position. To resist, we skipped the planes that were marked on us. We familiarized ourselves with details and points on our faces that were disregarded, forgotten, manipulated, or hidden. What was left were rhizomatic connections that connected one forgotten story with another, marking us by the multiple connections within ourselves and in relation to the Other.[9]

Someone else who became othered by the X-ray was the French scientist Prosper-René Blondlot. Blondlot had already discovered how to measure radio waves, but this was a small accomplishment in comparison to Röntgen, who in the late 19th century had achieved stardom. Determined to upstage Röntgen and assert himself as France's pre-eminent scientific pioneer, Blondlot worked unremittingly on novel X-ray-like devices. In 1903, in a dimly lit laboratory at the University of Nancy in France, he pulled the lever on his newly built device, causing a flicker of electricity to ignite and release a faint scent of ozone into the room. A subtle but detectable radiation emerged from his apparatus, filling the tightly packed space. Excited by this remarkable discovery, Blondlot diligently recorded his findings.

Blondlot believed he had stumbled upon a mysterious electromagnetic phenomenon that could challenge the latest scientific breakthrough in radiation. He named his discovery the "N-ray."[10]

According to Blondlot, his experiments revealed that N-rays emanated from various substances, including the human body, although curiously not from freshly cut wood. These N-rays possessed a faint phosphorescence and could be influenced by magnetic fields and prisms. Blondlot argued that the wave-like properties of N-rays allowed them to be stored in brick and masonry, facilitating their easy transportation. Like X-rays, N-rays could be absorbed or transmitted by different materials to varying degrees, potentially enabling the differentiation of tissues and structures within the human body, going beyond the bone as the standard for precise measurements of the internal record. Unlike the hazardous X-rays, however, N-rays could penetrate solid objects harmlessly, suggesting they could revolutionize scientific and medical practices. The news of this groundbreaking finding excited France, as it aimed to surpass Germany in the realm of scientific discoveries that delved into the inner workings of the human body. Yet shortly afterwards, Blondlot's discovery

[9] Cf. Édouard Glissant, *Poetics of Relation* (Ann Arbor, MI: University of Michigan Press, 1997).

[10] R. (René) Blondlot, *"N" Rays [Microform] : A Collection of Papers Communicated to the Academy of Sciences: With Additional Notes and Instructions for the Construction of Phosphorescent Screens*, trans. Julien François William Garcin (London, New York: Longmans, Green & Co., 1905).

faced scrutiny from the scientific community and was eventually discredited by American physicist Robert W. Wood.[11] But Blondlot yearned to discover so much that he continued to insist that he had discovered the N-ray. He insisted that he saw what nobody else could see, not only once but over and over again; the desire to discover shaped his vision. Blondlot, reveals the power of imagination and desire, and how what we see is intimately entangled with imagination and ideals. The physical anthropologists wanted to see the correlation between craniofacial measurements and craniofacial capacities, and Blondlot wanted to discover a different ray that would surpass Röntgen's.

In the same era, 1905, sociologist and historian William Edward Burghardt Du Bois published his short story *The Princess Steel*, wherein he characterized an innovative technology that surpassed the capabilities of X-rays and N-rays. His device, known as the Megascope, possessed extraordinary powers of observation. The Megascope could perceive not only the physical world but also delve into the spiritual realm and even transcend into the enigmatic fourth dimension of time. Du Bois' detailed description of his Megascope's capacity to perceive a temporal history of steel is as follows:

> Now the Great Near! And that problem I have solved by the microscope megascope, ... it is the Curve of Steel—the sum of all the facts and quantities and times and lives that go to make Steel, that skeleton of the Modern World. We will look through here and if all is well behold the Over-world of steel and its Over-men.[12]

Blondlot's N-rays and Du Bois's Megascope serve as both technological tools and literary devices, but their true power lies in their speculative nature and their potential to reshape social narratives. Speculation, designers Anthony Dunne and Fiona Raby argue, offers possibilities from social engagement, satire, and raising awareness; to inspiration, reflection, highbrow entertainment, aesthetic explorations, speculation about possible futures, and as a catalyst for change. Through speculation, technological narratives and devices

[11] Irving M. Klotz, "The N-Ray Affair", *Scientific American* 242, no. 5 (1980), 168–75.

[12] W. E. B. (William Edward Burghardt) Du Bois, *The Princess Steel*, 1905. W. E. B. Du Bois Papers (MS 312). Special Collections and University Archives, University of Massachusetts Amherst Libraries, 4.

can transcend the established rules and social constructs of our universe, allowing for a re-imagining of our technological and social expectations. By introducing new elements and alternative perspectives that were previously unimaginable within our existing worldview, they open ways of seeing alternate realms of possibilities. Merely accepting existing configurations of technology and humanity as sacred is insufficient. Speculation functions as a means of investigation and critique, challenging and questioning the deterministic perspective of technology and the limitations it imposes on individuals.[13]

Fold IV: Listening

While the details of how N-ray and Megascope technologies operated have been lost to history, in recent years we have successfully gathered enough information to create a unique apparatus that merges the penetrating abilities of N-rays with the temporal telescopic features of the Megascope. We have named this creation the MegaN-rayScope R-5000.

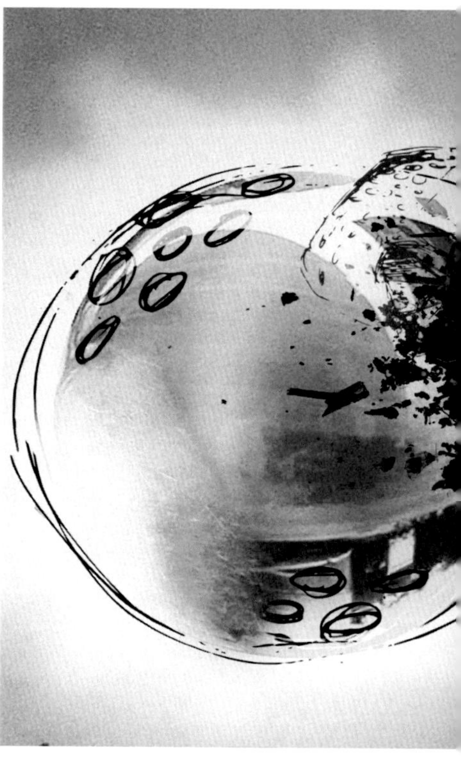

Fig. 2 – *R-5000 Seeing Itself.* Courtesy of the authors.

The MegaN-rayScope R-5000, (R-5000 for brevity) stands as a testament to human ingenuity. Forged from a harmonious fusion of advanced materials—compost, seawater for cooling, steel, and bioplastic—this prodigious creation exemplifies not only unwavering durability but also a deep-rooted commitment to environmental sustainability. At its very core resides a computational system of extraordinary dimensions, an ensemble of three integral elements: slime mould, Raspberry Pi, and skin cells. Slime mould breathes life into the device and is used to identify and navigate the most efficient computational pathways. In concert with the Raspberry Pi, a formidable single-board open-source computer, we were able to build the system into a palm-sized device that takes critical tasks of data analysis, storage and control to a new technological level. Simple, tractable, and with the computing capacity of a machine learning data farm. Skin cells cover the device,

[13] Anthony Dunne, and Fiona Raby, *Speculative Everything: Design, Fiction, and Social Dreaming* (Cambridge, MA: The MIT Press, 2013), 3.

which respond to the cells of the user, reducing time and eventually miscommunication between brain signals and motoneuron activity. Yet, the beauty of the R-5000 lies in its profound ability to transgress solid boundaries, to go beyond the fixed dimensions of the person in front of them to acknowledge and listen to their songs, stories, and movements.

The R-5000 possesses the remarkable ability to perceive and comprehend various sources, including devices, humans, animals, landscapes, and more. To showcase its capability, we affixed an assemblage of electrodes to the apparatus and implanted them into the earth. Setting the dial to its maximum level, we patiently awaited a familiar tremor. Diverging from conventional ground penetrating radar, our device forsakes a visual depiction of the subterranean realm in favour of an olfactory experience. Rather than perceiving in a linear manner, it navigates through undulating pathways to analyse the composition of soil layers, sometimes cracking kernels of sediment that have embodied smells of the past. Through the scope, we were able to discern the scents emitted by each stratum, unveiling a profound journey through time. The initial aromas were from the late 1990s and carried a pungent flavour of smoke. We winced, coughed, and gagged on the prickling, suffocating air. Bold letters in red appeared on the display: Indonesia forest fires, 1997–1998. We winced whilst the distant memories immersed us. Soon enough, we detected the scent of soot emanating from the 1968 Holy Week Uprising. Delving deeper, we encountered the essence of vast buffalo herds, just prior to their devastating decimation. Further still, as we turned the dial to encompass the western part of the African continent, we were greeted by the aroma of smoked fish, a testament to precolonial coastal settlements. Another turn and we sensed the verdant soil of the Mongolian steppes torn up by galloping horsemen. As we continued our descent, it transported us even further back in time, unearthing hidden layers of history until our senses were satiated.

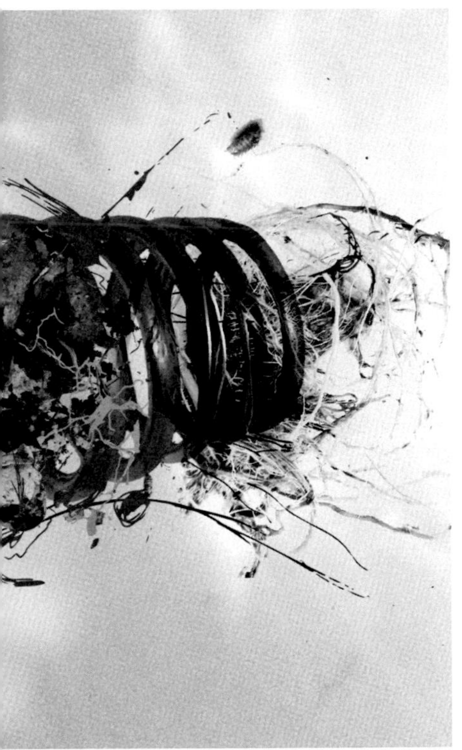

Then using ourselves as subjects, we directed the R-5000 towards our own bodies. This time, we were not entrapped by the cephalometer. We were softly moving, stretching our aching bodies. We pulled the cold slimy lever of the R-5000. With each deep breath of the machine, electricity pulsed through the wires and a great bright silent wheel whirled steadily, its cogs caught a black ball and sent it whirling till it stopped in the faint tinkle of a silvery bell.[14] It generated a steady high-pitched sound that was interrupted only by an unpredicted *sotto voce*. After which a fluorescent glow emanated from the solid crystal globe and cast a shadow behind us, bursting into song, welcoming those within the bones and flesh that have been rendered invisible, invisibilized. "Hello, how are you?"[15] The R-5000 asks the ancestors and spirits who shaped our pasts, who need to be named, and who are still in the present. We sense the invisibilized differently, some hear their answers through songs and others through the wind. The R-5000 catches their stories anyway. After sensing the persons and beings who shaped your past, the R-5000 listens intently to your body between its own breathing. It listens for the stories that remain within your healed bone fractures. It asks, in song, how you felt and lets the fractures and the scars speak. It asks why you ended up under the medical gaze and why you were hurt. When it looks at the fractures, it tracks the steps that you took, were forced to take and couldn't take. It listens to your intersectional harm, knowing that behind every image is a story that deserves to be told.

Finally, we placed our device in front of Röntgen's X-ray and again the globe ignited a green ray that cast shadows behind the X-ray machine. The X-ray machine, traditionally considered objective, now revealed a mess of curves and hesitations, a series of uncertainties that had once been framed as certainties. Wilhelm Röntgen's wife, whose X-rayed hand was displayed as a romantic gesture, spent lonely nights staring at the image when Röntgen was out displaying his findings. They were both in the X-ray's captivity, defined by its narrow conception of what it meant to be human. Inside the X-ray we could see its computational logic, the stories that had been lost, the debris of success and certainty.[16]

[14] As inspired by Du Bois's Megascope. Du Bois, *The Princess Steel*, 4.

[15] Documentary Film Institute, *Undisciplining the Fields: Abou Farman, with Tonya M. Foster*, 2023. https://www.youtube.com/watch?v=3W65LK0U430. Thank you, Abou Farman, for raising this question.

[16] For our intrepid readers, we recorded some of our device's songs for your listening pleasure: www.ayo.io/R-5000/

Fold V: Unfolding

And then with a great thunderous clap, the machine crashed releasing billows of vibrant green ether that filled the firmament. With each burst of energy, it released a spark and a unique sound, conveying a liberation from the linear boundaries of the X-ray object, releasing itself from the one-dimensional narrative that had imagined it. X-rays, as with all technologies, are products of a certain disposition, a way of looking at, reading, and writing the world, and thus they are literary devices as well as technological ones. In this essay we have read about the inventors and the early adopters of the X-ray, speculating what they sensed and desired. We have danced to resist the X-ray, we have imagined ourselves to be a scientist who so deeply wanted to be recognized for their findings that he ended up seeing a new phenomenon, and we contrasted his 'technology' with a fictive technology from Du Bois. Finally, we invented our own technological device as a response to the linear and one-dimensional way of seeing that the X-ray advocates. By speculating otherwise, we recognize the importance of critically engaging with technologies to construct our futurity, one that embraces a multiplicity of narratives and ways of telling stories. By speculating about technology, we also change it. Like when folding a piece of paper several times, traces of the folding endure, can be read, and reinterpreted when the paper is unfolded.

References

Blondlot, R. (René). *"N" Rays [Microform] : A Collection of Papers Communicated to the Academy of Sciences: With Additional Notes and Instructions for the Construction of Phosphorescent Screens.* Translated by Julien François William Garcin. London, New York: Longmans, Green & Co., 1905. http://archive.org/details/nrayscollectiono00blonrich.

Broadbent, B. Holly. "A NEW X-RAY TECHNIQUE and ITS APPLICATION TO ORTHODONTIA". *The Angle Orthodontist* 1, no. 2 (1 April 1931), 45–66.

——— "The Face of the Normal Child". *The Angle Orthodontist* 7, no. 4 (1 October 1937), 183–208.

Daston, Lorraine, and Peter Galison. *Objectivity*. Princeton, NJ: Princeton University Press, 2021.

Documentary Film Institute. *Undisciplining the Fields: Abou Farman, with Tonya M. Foster*, 2023. https://www.youtube.com/watch?v=3W65LK0U430.

Du Bois, W. E. B. "The Princess Steel," in Brown, A, and Rusert, B. "Introduction" in *PMLA* 130, no. 3 (2015), 819–829.

Dunne, Anthony, and Fiona Raby. *Speculative Everything: Design, Fiction, and Social Dreaming.* Cambridge, MA: The MIT Press, 2013.

Glissant, Édouard. *Poetics of Relation*. Ann Arbor, MI: University of Michigan Press, 1997.

Klotz, Irving M. "The N-Ray Affair". *Scientific American* 242, no. 5 (1980), 168–75.

Krogman, Wilton Marion. 'Contributions of T. Wingate Todd to Anatomy and Physical Anthropology'. *American Journal of Physical Anthropology* 25, no. 2 (1939), 145–86.

"MegaN-RayScope 5000," 13 Oct 2023, http://www.ayo.io/R-5000/

Morris-Reich, Amos. "Anthropology, Standardization and Measurement: Rudolf Martin and Anthropometric Photography". *The British Journal for the History of Science* 46, no. 3 (2013), 487–516.

Mould, R. F. *A Century of X-Rays and Radioactivity in Medicine: With Emphasis on Photographic Records of the Early Years*. Boca Raton, FL: CRC Press, 1993.

Sekula, Allan. "The Body and the Archive". *October* 39 (Winter 1986), 3–64.

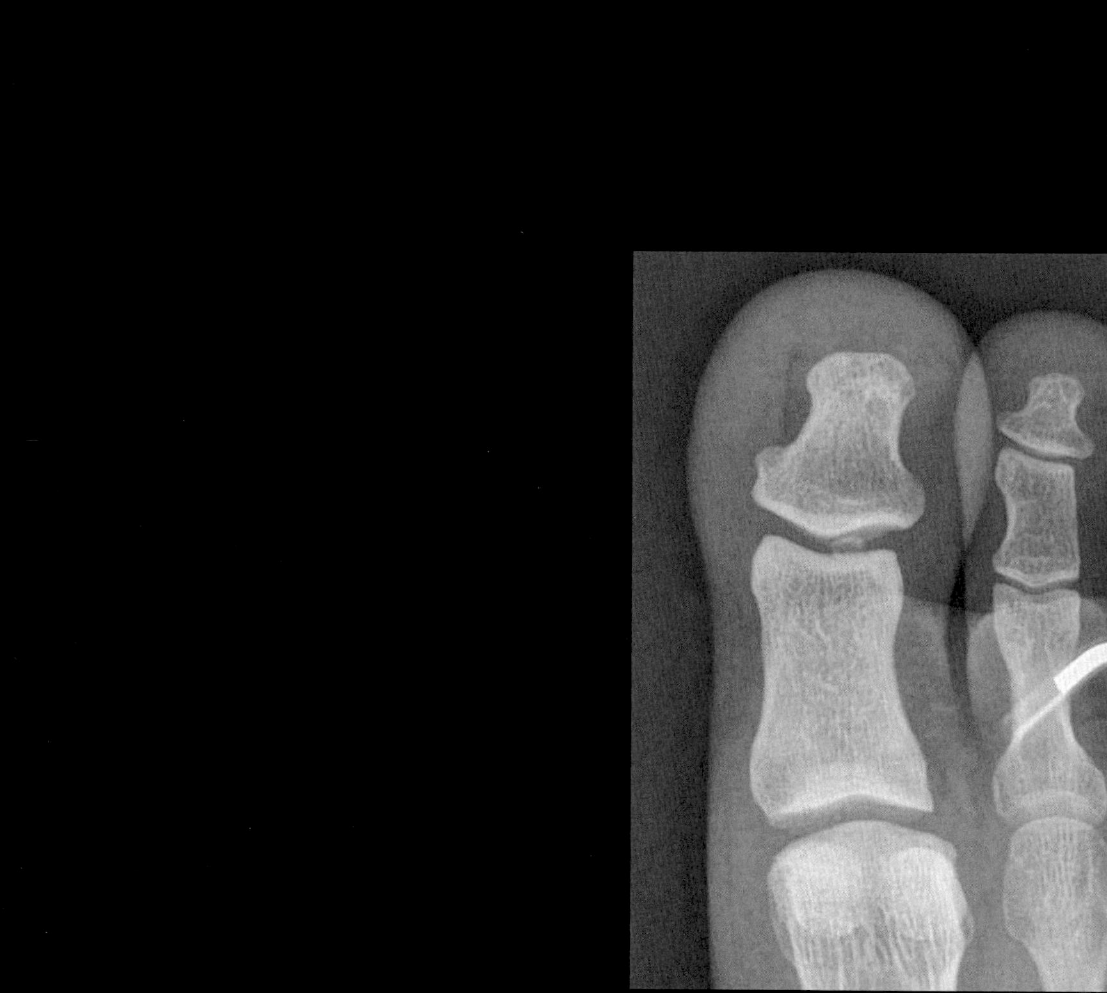

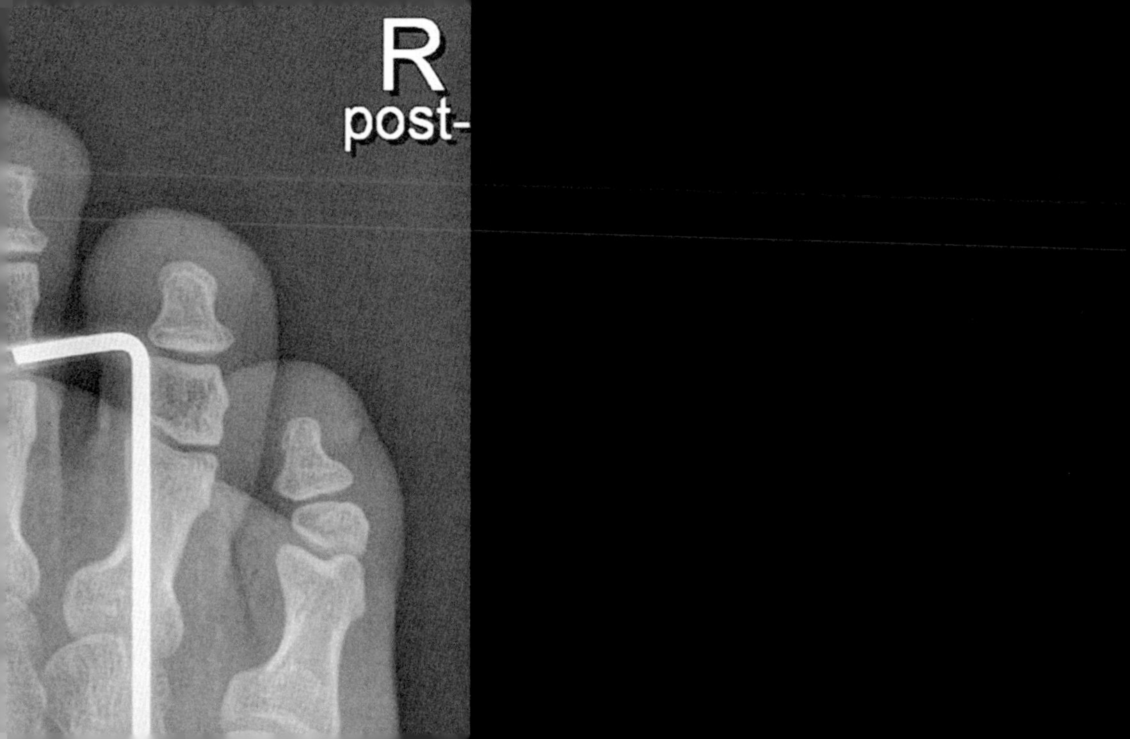
R
post-

Paolo Favero

Deep Visions and Superficial Desires: X-rays as Science and Popular Culture

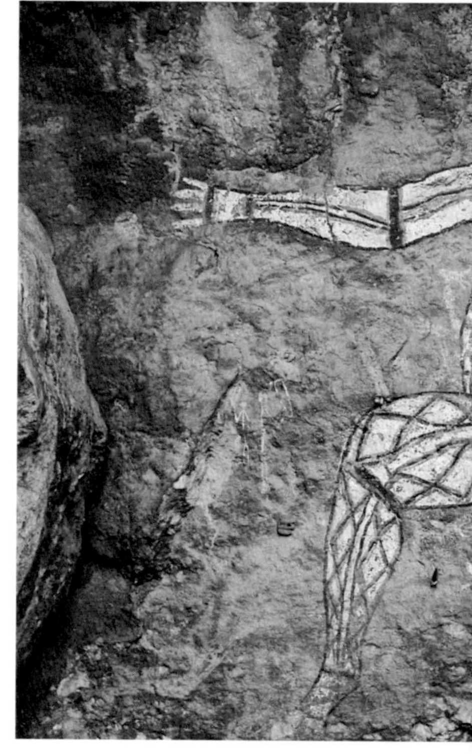

Fig. 1 — Aboriginal X-Ray Style Rock Art, Anbangbang Rock Shelter, Kakadu National Park, Australia. Author: Thomas Schoch. 2005. The image is licensed under the Creative Commons Attribution-Share Alike 3.0 Unported. https://en.wikipedia.org/wiki/X-ray_style_art#/media/File:Aboriginal_Art_Australia(3).jpg.

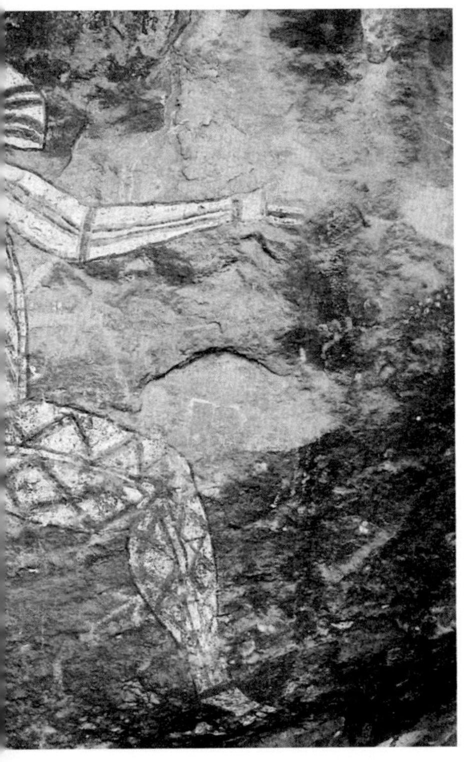

Reflecting on X-ray imagery, the first thought goes to serious matters. Deep visions are, after all, a matter of deep truths. Many of us have, at some point in life, had an accident or a check-up that required an X-ray image to be produced. And most of us have trusted the machine's capacity to produce a factual insight, situated beyond our perceptual capacities. Our thoughts may also wander to X-ray images in the context of surveillance, those used for preventing the movement of people and goods across borders. Again, deep visions for deep truths; matters of life and death. But there is also another, more superficial and perhaps morally dubious side to X-ray imagery. This is a space of hidden desires and sexual fantasies. A space of popular culture where deep vision stands for superficial arousal. A space with its own particular politics. This space is the object of my confabulations.

In the 1970s a series of advertisements circulated prominently in popular magazines and newspapers in Europe and elsewhere. It marketed "X-ray glasses" (or "X-ray specs", or "gogs"); a commercial item that promised to make deep vision cheaply available for anyone. The item rapidly became the secret object of desire for heterosexual male buyers. Promising the power to see beyond the barriers of materiality, these specs, as visible in the ads, stimulated the fantasy of seeing beneath the clothes of women (and also, but that was secondary, through your own skin). Sold by mail-order companies, these specs were often ordered in secrecy. They were pitched at, teased and reinforced the erotic fantasies of heterosexual male audiences.

The X-ray glasses were, needless to say, a hoax. Consisting of a double layer of cardboard "lenses" with a hole in the middle with a small feather (or lens) diffracting the light, the glasses simply showed the wearer offset images. Yet these images reproduced the uncanny aesthetics of X-rays with their double layer made up of a darker body (a silhouette) and a surrounding shadow image. With their cheeky, unethical and perhaps immoral nature, X-ray specs point us, however, in the direction of a dialogue between the search for knowledge and for pleasure. The modern desire to see beyond materiality (the deep vision promised by X-rays) is inevitably connected to the ruthless modern desire and pleasure for penetrating, digging, excavating for, and extracting truth (metaphors and practices that are today finally under attack in the ruling debates on the decolonization of science and popular culture).

Before I go any further, I want to ask the reader to briefly stop and meditate upon the sheer beauty of the X-ray picture. This is a special kind of image that belongs simultaneously to different image types. If we are to follow W. J. T. Mitchell's 1986 typology of the family of images, we can say that X-ray pictures are both "graphic" and "optical."[1] According to Mitchell, the former refers to visible objects like photographs, sculptures, or drawings; the latter to those images generated within the realm of natural phenomena (a mirror, a projection). The X-ray image is a combination of both. It is obviously graphic (it is an image-object) but it also looks like a shadow. Like a shadow, it is ambiguous; it has unclear boundaries and is capable of overlapping a vast quantity of layered data on its two-dimensional surface. Like shadows, X-ray images seem to play with the affirmative nature of negation. An X-ray image functions according to the same visual principles that guide the design of Sufi Mausolea, where the divine (and therefore, meaning) emanates visually in the play of shadows and light that are created by the beautiful marble grids. Similarly, X-ray images, too, produce meaning through the contrast between light and darkness as it appears on the light box. Conventionally viewed in negative, they leave to the dark parts the duty to display the light (organs appear dark in X-ray negatives). X-ray images are hence existential objects, and they remind us of the uncertain continuum of darkness and light that makes up life.

[1] W. J. T. Mitchell, *Iconology: Image, Text, Ideology* (Chicago, IL: University of Chicago Press, 1986).

The visuality of X-ray images reminds us of many other works of art. Think of the similitude with the so-called X-ray style art of ancient Aboriginal Australian populations (approximately 3000 to 4000 years BCE); or of the mysterious visions offered by the Shroud of Turin. Pushing this connection even further, think of the so-called *Anatomical Machines* contained in the basement of the Sansevero Chapel in Naples, Italy. Testifying to the passion for science and art of the Prince of Sansevero and to his longing for seeing bodies from the inside, these are two skeletons (a man and a pregnant woman) displaying a system of red and blue arteries and veins. Made of beeswax, iron and silk, these bodies were long considered the result of magical or alchemic intervention. The same suspicion was also raised with regard to Giuseppe Sanmartino's *Veiled Christ* (1753), a depiction of a dead man (Christ) lying, whose every muscle and expression are visible under what looks like a thin veil of wet gauze. Rumour has it that upon delivery, Sanmartino was suspected of having embalmed a real human being and solidified a real thin veil in order to obtain this effect. This statue, along with other sculptures that surround it, defies the impenetrability of materials and displays, in a game of veiling and unveiling, a tense play of eroticism and thanatology.

There have been many attempts at penetrating the skin of bodies and defying the materiality of nature. X-rays are just one more step in that journey. And this is a journey, as witnessed by the case of the Sansevero Chapel, that is also filled with dangers. Indeed, we do know that X-rays are physically dangerous for the 'imaged' subject; but they are also dangerous at a symbolic level, representing a transgressive desire to penetrate reality, to defy common perception, conventions and 'normality.'

It is obvious then, that X-ray vision would also become an ability of superheroes. The capacity to see in X-ray appeared for the first time in the 1930s, embodied by female comic book superhero Olga Mesmer, who had developed a capacity to see behind walls due to exposure to X-ray radiation as a child. This quality inspired the creators of Superman, the hero who turned X-ray vision into a weapon to protect humans. Defending life from evil powers, Superman however, falls prey to the erotic temptations offered by this medium. In *Superman: The Movie* (1978) he inspects not only Lois Lane's lungs but also her underwear. James Bond too, who has X-ray glasses as a part of his 'spy gadgetry,' does something similar in

The World is Not Enough (1999). During a mission to identify concealed weapons at a casino, he cannot resist the temptation of glancing beneath the clothes of female staff members. The X-specs with which I opened my confabulations surely gathered inspiration from these scenes.

Popular culture added to the X-ray fantasies of deep vision, those of invisibility (an association born of the capacity of X-rays to make bodies transparent). In Jules Verne's *The Secret of Wilhelm Storitz* (written just a couple of years after the first X-ray image taken by Wilhelm Röntgen in 1895 and published in English only in 1963) the central character learns to make himself invisible and throws a whole city into panic. Put into perspective with Verne's other novels, this book offers an evident insight into the connection between (in) visibility, knowledge and depth. The novel constitutes a natural prolongation of Verne's interest for science and technology that in other novels such as his 1864 *Journey to the Center of the Earth* (the first English translation appeared in 1871) and the 1870 *Twenty Thousand Leagues Under the Sea* (appearing in English in 1873) are metaphorized through the idea of physical depth. The world of erotic comic books too is populated by invisible men. In Milo Manara's *Butterscotch: The Flavor of the Invisible* (1989) a man who has discovered an ointment capable of making him unseeable, invades the intimacy of a young woman and starts an erotic relationship with her thanks to his invisibility.

Popular culture has not only played with the pleasures of deep vision but also with the dangers attached to it. In the 1963 movie *X: The Man with the X-ray Eyes* Dr. Xavier is the inventor of a liquid that when dropped into the user's eye expands the spectrum of vision into the realm of ultraviolet and X-rays. Like a superhero, Dr. Xavier starts using his power for noble purposes. Very soon, however, he loses control of the drug and begins a descent into crime. The capacity to see through things quickly becomes a tool for making money. A miracle diagnostician and a casino cheater, Dr. Xavier ends up escaping to the desert where, blinded by his powers, he eventually meets an evangelist who, seeing in his deep vision the doing of the devil, invites him to pluck out his eyes.

Fig. 2 — Theatrical release poster of *X: The Man with the X-ray Eyes* by Reynold Brown, 1963.
The work of art itself is in the public domain.

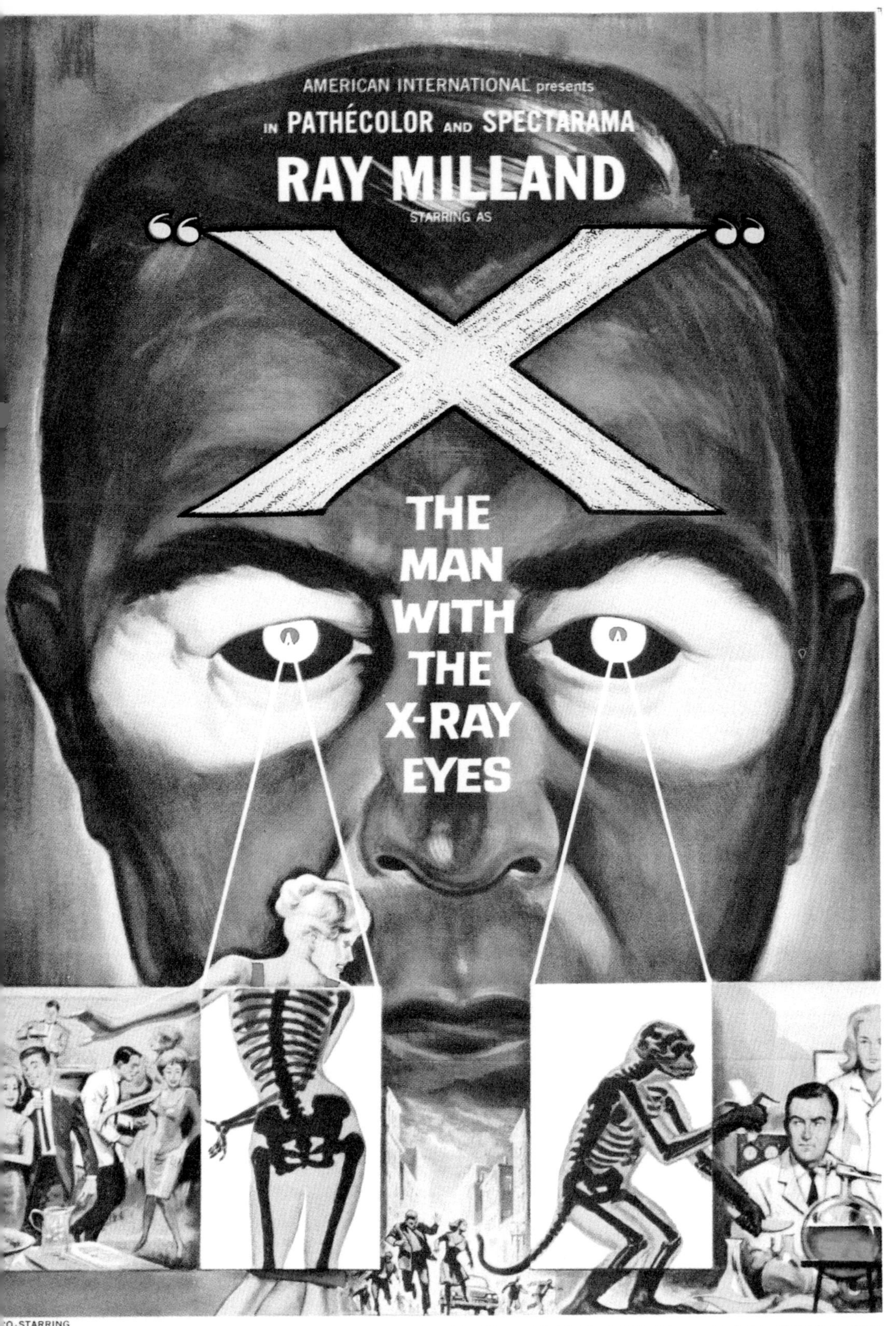
AMERICAN INTERNATIONAL presents
IN PATHÉCOLOR AND SPECTARAMA
RAY MILLAND
STARRING AS
"X"
THE MAN WITH THE X-RAY EYES

CO-STARRING
DIANA van der VLIS · HAROLD J. STONE · JOHN HOYT and DON RICKLES
Produced and Directed by ROGER CORMAN · Screenplay by ROBERT DILLON and RAY RUSSELL · Story by RAY RUSSELL · Executive Producers: JAMES H. NICHOLSON and SAMUEL Z. ARKOFF · Music by LES BAXTER

In Western popular culture X-ray vision is represented as a tool for transgressing the boundaries and limits of neurotypicality, of normality and morality. This urge to move beyond materiality and the sensory capabilities of our bodies is however, also something dangerous. It defies the border that separates life from death. Producing predictive visions of the body as it will one day look like (i.e., a skeleton), X-ray images break a taboo. They anticipate and visualize our death. As rumours have it, when Röntgen's wife saw the X-ray image of her hand, she exclaimed that she had seen her own death. This predictive capacity has been the object of attention in popular culture and art. Artist Nick Veasey consciously plays with this idea in his X-ray portraits, transforming the most mundane poses and situations into extraordinary actions performed by what looks like skeletons. Life and death meet in the space of Veasey's images testifying that there is something sublime to X-ray imagery. These images come very close to death, yet they never fully enter into contact with it. They witness it from a place of safety, like a Kantian observer would witness a mountain avalanche the from the safety of a warm hut. X-ray images are sublime objects that consolidate the nexus between photography and death that has been central to the history of this medium.[2] And they take this connection further.

The photography of Lennart Nilsson surely consolidates the yearning for going beyond the surface of the human body. Showing it from the inside for the first time using an endoscopic camera, Nilsson created another space where science and popular culture meet. The 1965 close-ups of a foetus (*Foetus 18 Weeks*) created a degree of turmoil in society, also polarising the debate on abortion. Inner pornography takes on a similar endeavour today, also deploying the same kind of cameras. Inverting established points of observation, it conventionally shows sexual intercourse from the inner perspective of an organ being penetrated. Today, hyperspectral cameras offer probably the most (subtle) expression of this longing. Using both visible light and invisible near-infrared light they simultaneously stay above and enter beneath the surfaces of objects. Producing patterns capable of revealing what human eyes are missing, they can, among other things, show the skin and structure of a hand

[2] Cf. Roland Barthes, *Camera Lucida: Reflections on Photography*, trans. Richard Howard (London: Cape, 1982).

together with the veins beneath it. Hyperspectral cameras offer perhaps a new 'hybrid vision' capable of simultaneously contemplating the inner and the outer, the deep and the superficial. This would be a very much needed non-dualistic shift in discourse.

In her book *Good Looking* (1996) Barbara Maria Stafford enquires into the Western distrust for surfaces.[3] A key characteristic of Western culture, she suggests, is the reduction of the perception of reality to a matter of language and hence, I may add, of thought. Is this what lures behind this yearning for and fascination with deep vision? A celebration of the (Cartesian) separation of body from mind? Of the supremacy of the mind and the soul and the (possibly erroneous) identification of the mind/soul as an 'inner' quality? Contemporary neurosciences are today enquiring into the latter aspect. Having identified the locus for practically all types of human perception they can still, however, not identify where the sense of self (or consciousness) resides. Possibly, exactly because it is not 'inside'? Neuroscientist Chris Niebauer asks "What if the brain is connected to, or a part of, consciousness—rather than a possessor of consciousness?"[4]

X-ray images and vision take us to these reflections, because they are much more than a pure matter of (hard) science. They are part of a much broader narrative and represent a yearning, that for deep vision, that is simultaneously also a question of (popular) culture. In this desire for going beneath the surface of the skin, for penetrating the human body and the world, and for overcoming the limits of neurotypical perception, the search for knowledge coexists with that for pleasure and arousal. This is a terrain of 'visual culture', a space where scientific discourse meets cultural narratives; where the quest for advancement meets the risks of extractivism; where sexual fantasies (and their attached forms of masculinity) meet an existential quest for exploring the thin border that separates life from death. Here Eros meets Thanatos and superficial desires prove to be a matter of deep truth.

[3] Barbara Maria Stafford, *Good Looking: Essays on the Virtue of Images* (Cambridge, Mass: MIT Press, 1996).
[4] Chris Niebauer, *No Self, No Problem: How Neuropsychology Is Catching Up to Buddhism* (San Antonio, TX: Hierophant Publishing, 2019), 204.

References

Barthes, Roland. *Camera Lucida: Reflections on Photography.* Translated by Richard Howard. London: Cape, 1982.

Milo Manara. *Butterscotch: The Flavor of the Invisible.* New York: Nbm Pub Co, 1989.

Mitchell, W. J. T. *Iconology: Image, Text, Ideology*. Chicago, IL: University of Chicago Press, 1986.

Niebauer, Chris. *No Self, No Problem: How Neuropsychology Is Catching Up to Buddhism.* San Antonio, TX: Hierophant Publishing, 2019.

Stafford, Barbara Maria. *Good Looking: Essays on the Virtue of Images.* Cambridge, Mass: MIT Press, 1996.

Verne, Jules. *Journey to the Center of the Earth.* 1871.

Verne, Jules. *Twenty Thousand Leagues Under the Sea.* 1873.

Verne, Jules. *The Secret of Wilhelm Storitz.* 1963.

Films and Artworks

Apted, Michael (Director). *The World is Not Enough.* 1999.

Corman, Roger (Director). *X: The Man with the X-ray Eyes.* 1963

Donner, Richard (Director). *Superman: The Movie.* 1978.

Nilsson, Lennart (Photographer). *Foetus 18 Weeks.* 1965.

Sanmartino, Giuseppe (Sculptor). *Veiled Christ.* 1753.

Salerno, Giuseppe (Anatomist). *Anatomical Machines,* around year 1763.

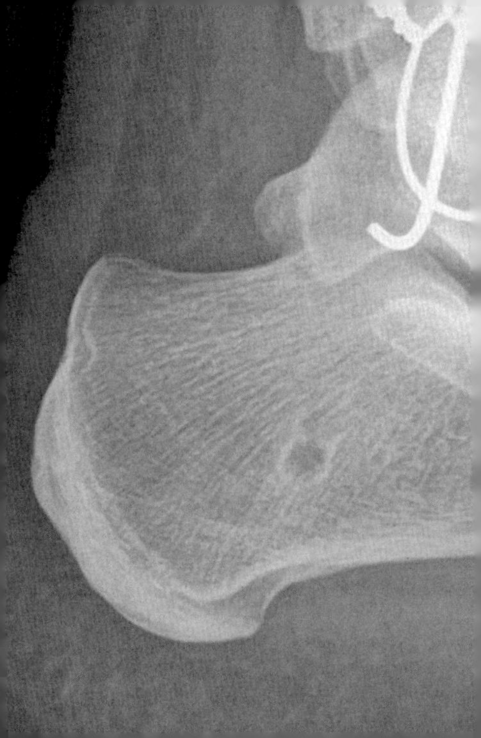

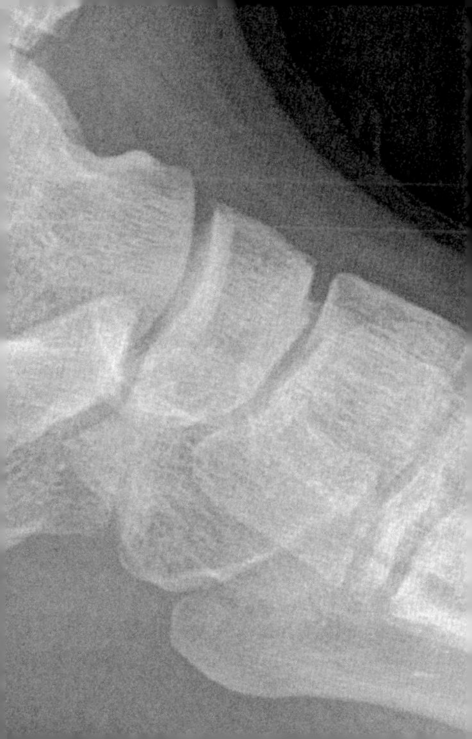

Sandra Noeth and Joanne 'Bob' Whalley

Bothering the X-ray

I'm Sorry

I'm sorry to bother you, as I try to find the right words to say hello and goodbye to you through the inky texture beneath my fingers. As I struggle to find the right words in and amongst the medical discourse and the imaging technologies of the X-ray, I struggle with my articulation, and I find myself wanting. It feels important to me, at the beginning of this bothering, to attempt a formation of words. To find words to describe the juncture between my body and a medical discourse which casts its eyes over me. A medical gaze which bypasses the warmth of my flesh to insist on practices which compartmentalize and fragment a body. The X-ray is part of this pretence. In an attempt to feel at the corners of the idea of the X-ray, I suspect I am an ill body, bookmarking a place in a system of anatomy and diagnosis.

Looking Through

Why should our bodies end at the skin?

I am feeling how sweat collects in my pores as I am waiting for the results of the medical examination. A check-up. Routine. No cause for concern. I distract myself and concentrate on my skin. It often becomes transparent on the X-ray images, like insect shells after their moult. Bones appear white, air appears black, and muscles and soft tissues appear grey. The skin is "not smooth, flat and neutral," writes Anne Juren, but a porous landscape with marks, hairs, folds, wrinkles, lines, holes, creases and traces that embody three-dimensional spatial and temporal dimensions. It moves, breathes, sweats, smells, and regenerates, changing constantly.[1]

It is the largest respiratory and sensory organ of the human body. Even when we stop breathing, or when our breath is stopped by others, incorporating relation and movement, offering a different kind of touch, the skin is open to the world and to others. The skin fades out when overexposed to light. It withdraws from the dark and blue surfaces of the digital image processing. In view of the medical gaze, it stays in the negative—while recollecting often small-scale acts of intrusion on the body, it insists on being present as well.

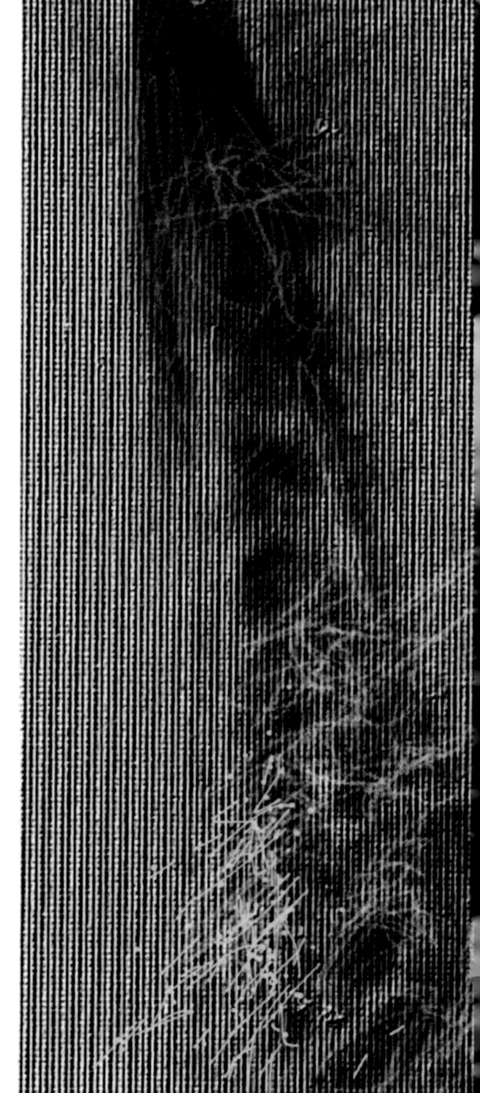

[1] Anne Juren, "Lesson on the Skin, Plastering the Body", in *What Does It Take to Cross a Border? A Reader* (Berlin: ifa-gallery, 2019), 9.

Lightbox drawings 1–5, 2023. Courtesy of Joanne 'Bob' Whalley.

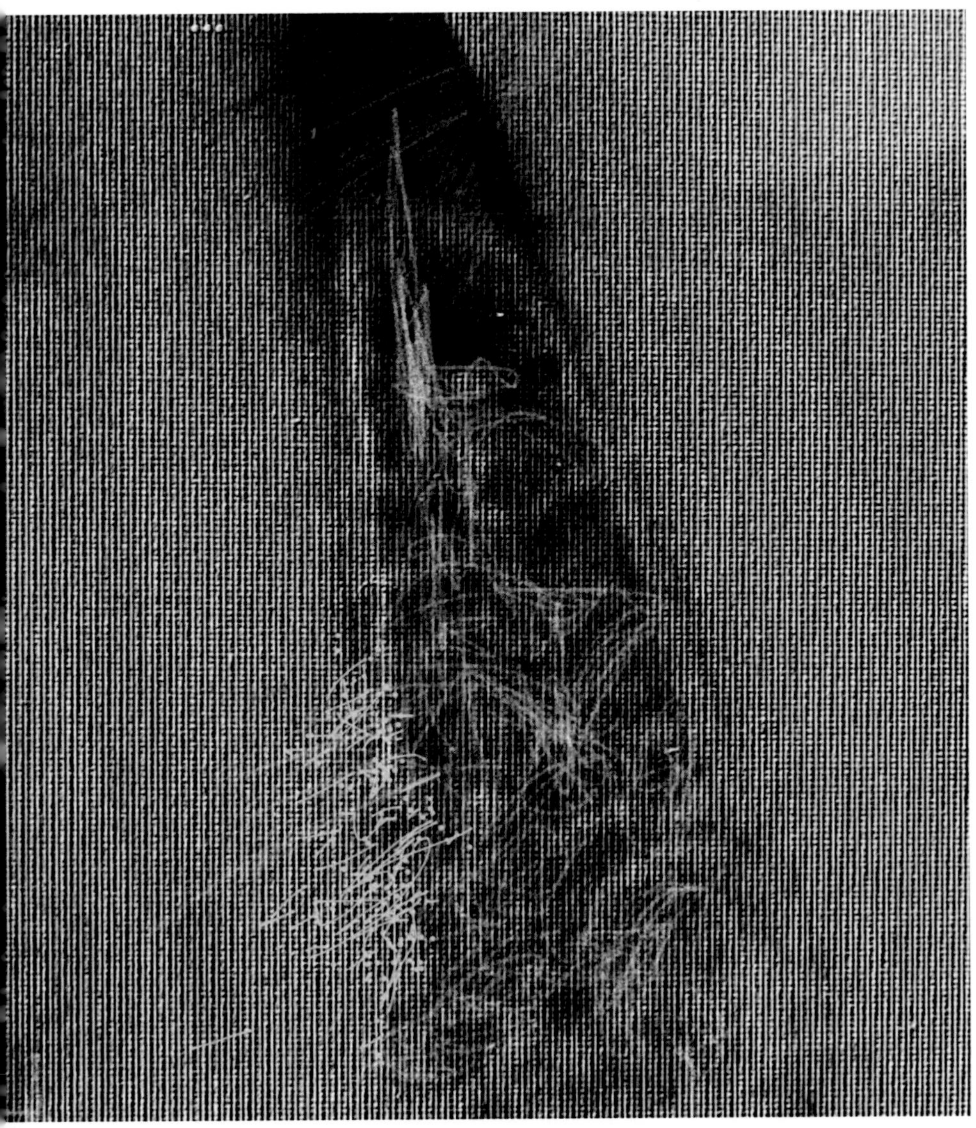

Inside-Outside

I want to ask you if you feel the same when you approach the X-ray. Do you slip inside the image, experiencing that body, your body? Or maybe the X-ray is just a picture, a quiet place where the collective noise of the collected bones lays down. Is an X-ray a kind of perfect storm? Everyone knows that at the heart of every storm there is a moment of perfect stillness, an eye from which we see things differently. In this pause, in this quiet place, perhaps I can perform how to be with an X-ray. I half close my eyes to the X-ray resting on my knees and observe the blurring between the inside of the body and the outside: a kind of "being singular plural."[2] I begin by half-closing my eyes, restricting the light entering the retina of my eye. My pupils, large and inky in response to the "blue-gray mist"[3] in front of me being converted into nerve impulses that shift and travel down my optic nerve. This downcast physicality, the lowering of my eyes, feels both polite and tentative: I am touching an assortment of body parts and this is an intimate act.

[2] Jean-Luc Nancy, *Being Singular Plural*, trans. Anne O'Byrne and Robert Richardson (Stanford: Stanford University Press, 2000), 1.

[3] John A. Reid, "The 'visibility' of x-rays", *Oral Surgery, Oral Medicine, Oral Pathology* 34, no. 2 (1 August 1972), 330.

The X-ray is strangely warm, not body-warm, but warm enough. I hold you with the lightest touch as I do not want my fingerprints to leave a trace. My fingers hover over acetate, the materiality and technology of the X-ray combining. This is a (body) practice which privileges soft liquids, what geographers Philip Steinberg and Kimberly Peters foreground as "wet ontology."[4] I understand, or rather sense, that when away from a warm body, in front of this image, we are many, singular and plural, all at the same time. In a moment of loss of a warm body, of loss of being with a warm body, the X-ray can acknowledge this dance traced in the corners of my vision. As I begin to sway, my half-closed X-ray gaze lets in shimmering colours, oil-like on the surface of your skin. I remember that the most commonly used reflective materials for X-ray mirrors are gold and iridium and lead, and I make myself laugh as I try, and fail, to imagine reflections which are simultaneously shiny and dull.

[4] Philip Steinberg, and Kimberley Peters, "Wet Ontologies, Fluid Spaces: Giving Depth to Volume through Oceanic Thinking", *Environment and Planning D: Society and Space* 33, no. 2 (1 April 2015), 248.

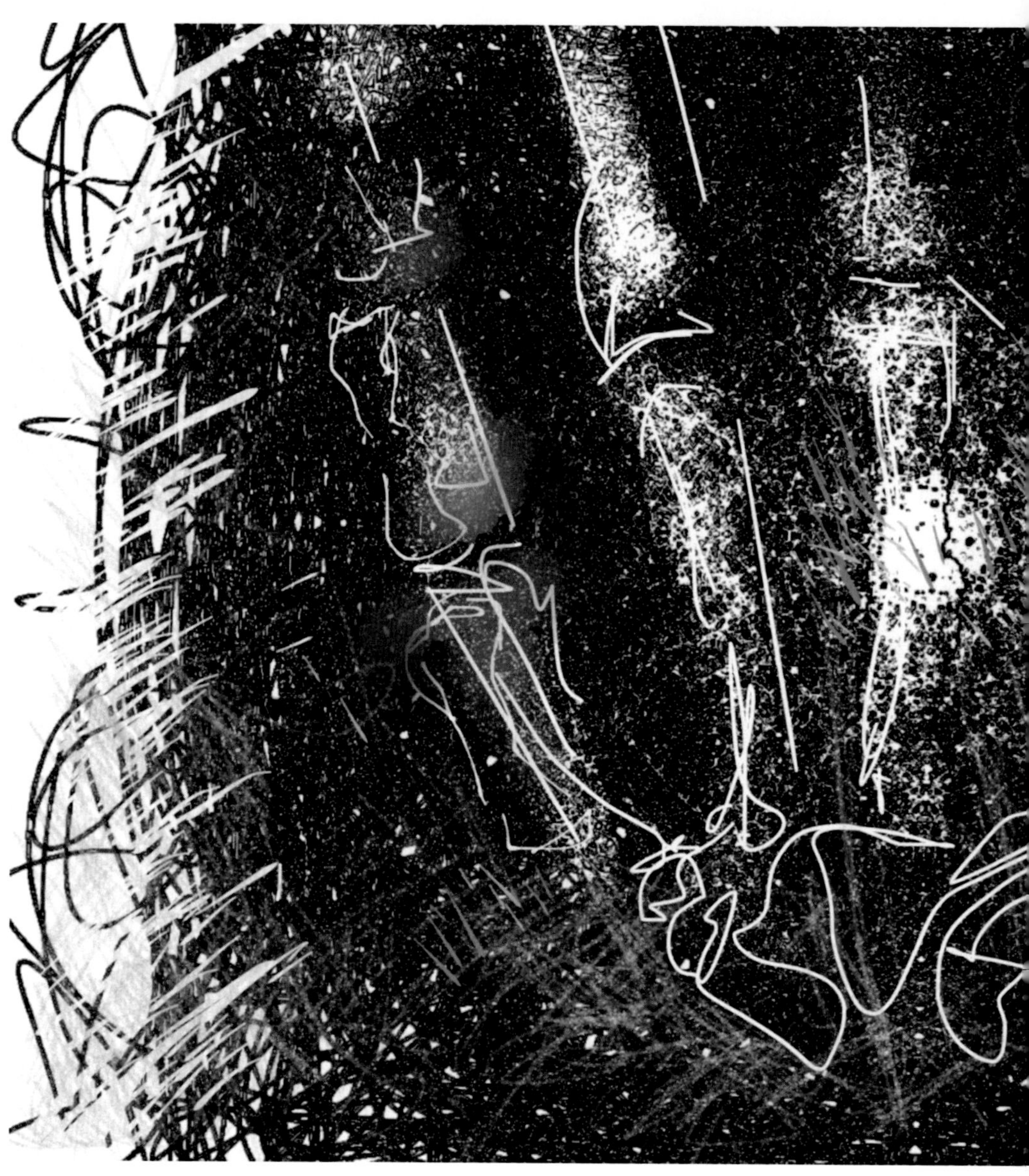

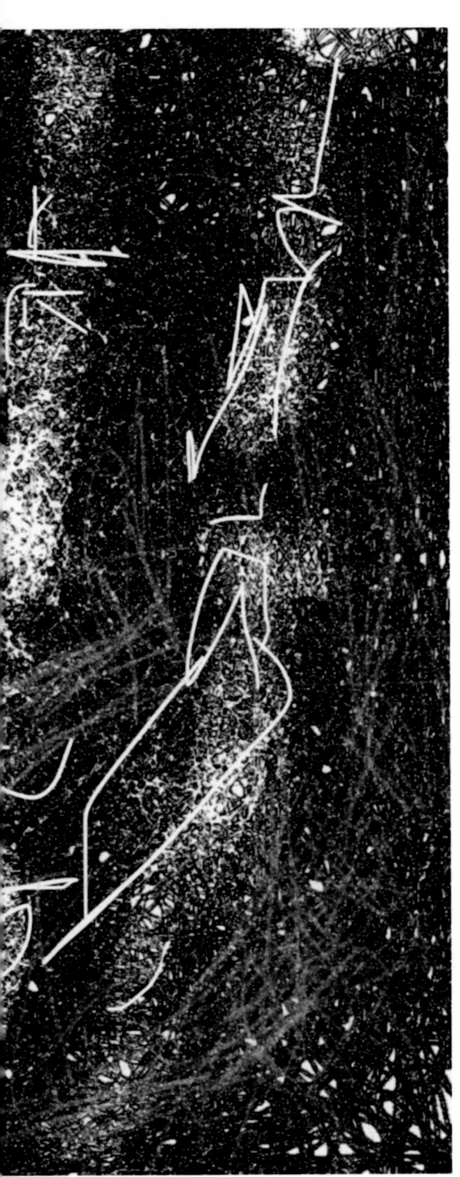

The Capacity of Having a Body

X-raying turns the body into a promise. A promise that there is something to be saved, a promise of wholeness and integrity. A promise of protection, as well: protection from physical intrusion against the material body and from attacks against its symbolic force. In many legal documents framing the right to bodily integrity in international humanitarian law, the promise of protection is linked to exactly this: a body that is your own, individual, with a name. A body that is distinguishable from other bodies. The capacity of having a body in this sense is connected to its capacity to belong—to belong to law, to medicine, to subjectivity. Yet, there is nothing to fix, as the body undermines the authority of technology and anatomy. Rendered public in the event of the X-ray and the political consequences of being over-exposed, the body turns us back to our phantasies and imagination—to a place of responsibility for the images and narratives that we create. A responsibility that recalls the X-ray as a specific memorial to a moment already passed, and also a device that conveys complex anxieties and feelings in the present.

Rotation

I am a body which benefits from rotation. Where the aspect of each clavicle is equidistant from the spinous processes. Where the spinous processes align vertically against the vertebral. Whereas

> [t]he face is a surface: facial traits, lines, wrinkles; long face, square face, triangular face; the face is a map, even when it is applied to and wraps a volume, even when it surrounds and borders cavities that are now no more than holes,[5]

as we feel below the surface of the skin, palpate a little, what do the "border cavities" of the X-ray offer the reader?
How can our consideration of X-rays through "fingery eyes"[6] understand these smooth and striated articulations of the body? What is this body? What would I like to see in this body even before I can see it?

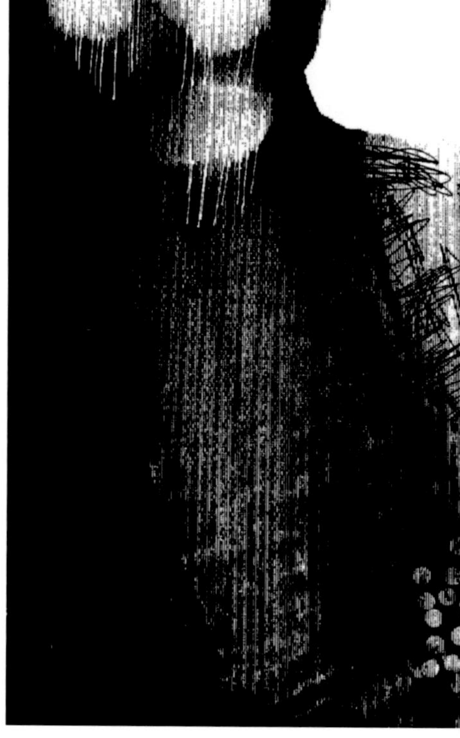

There is a manipulative force at play here, of the X-ray that manipulates light and sight at the same time: while it promises, and performs, narratives of transparency and care, it actually might distance us from encountering a body, our body.
Maybe I am a reader of the map, a map defining the body as a territory, one that is subject to different orders of possession—the law, medicine, subjectivity. If so, this is certainly a kind of counter-mapping, a subversive cartographic engagement. The manifold methods of looking at movement in bodies do not feel appropriate. What method shall I use to "pass through" an X-ray?

[5] Gilles Deleuze, and Felix Guattari, *A Thousand Plateaus: Capitalism and Schizophrenia*, trans. Brian Massumi, 2nd edition (Minneapolis: University of Minnesota Press, 1987), 170.
[6] Donna J. Haraway, *When Species Meet* (Minneapolis: University Of Minnesota Press, 2007), 5.

Perhaps a kind of looking which engages through "transforming loops."[7] Dancer-scholar Chrysa Parkinson was introduced to the concept of loops in choreographic practice when working with choreographer Mette Ingvartsen, and I wonder if this might be a method for a close reading of an X-ray. Neither a vertical way of looking, which moves from back to front, nor a sinistrodextral reading from left to right. Sara Ahmed suggests "[d]epending on which way one turns, different worlds might even come into view"[8] and I loop into my X-rays, softening any latitudinal and longitudinal grasp. The coordinate systems of medical technologies "may be understood as technologies of selfhood and embodiment, and use of technologies may engender certain kinds of identities,"[9] however the X-ray is also forever wed to the moving body, being a still promise to intense movement: a thrumming pulsation which lies through the acetate. I get lost for a moment in the layers of connections, and "what surrounds it, what formed it, what happened there, what will happen there."[10] This X-ray looking is a bordering dance where movement is stitched through grey bodies.

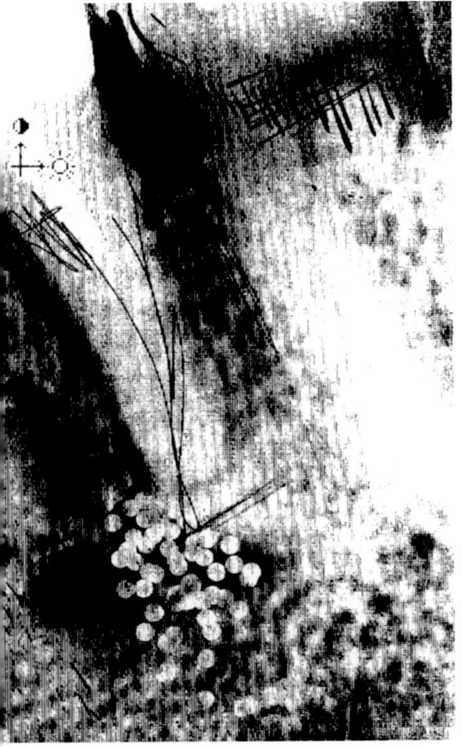

[7] Chrysa Parkinson, *walk + talk* (performance lecture, transcription, March 18, 2011).
[8] Sara Ahmed, *Queer Phenomenology* (Durham: Duke University Press, 2006), 15.
[9] Deborah Lupton, *Medicine as Culture: Illness, Disease and the Body*, 3rd ed. (London: Sage Publications Ltd, 2012), 46.
[10] Lucy R. Lippard, *The Lure of the Local: Senses of Place in a Multicentered Society* (New York, NY: The New Press, 1998), 7.

Inspiration

I am a body which encourages an in and out from an inspiration-exploration process.

I am a body where the lateral rib edges are visible. The X-ray is a window, a relation of inside to outside and outside to inside, depending on where you are kneeling. In her poem *Medusa*, Sylvia Plath is dismayed at being "overexposed like an X-ray," that her skin is a window that reveals everything.[11] When observing that other form of medical imaging, the sonogram, Karen Barad writes that "the sonogram has that 'x-ray' look to it: the body is rendered 'transparent' because a cross-sectional view does not privilege surfaces; in a sense, it has no respect for surfaces whatsoever."[12] And there are political implications of being over-exposed, of not being able to hide. The exhaustion. This is a body that is unable to be invisible.

I am on the threshold of seeing the body / a body / my body. The X-ray is the tracing of a warm line, of being here and there, inside / side / through a body. It can be playful being under the skin, to "be with" the ossified bones, to be nestled alongside a congested heart and lungs. The X-ray manifests itself with the turn of a hip and through the last glint of exhausted cartilage.

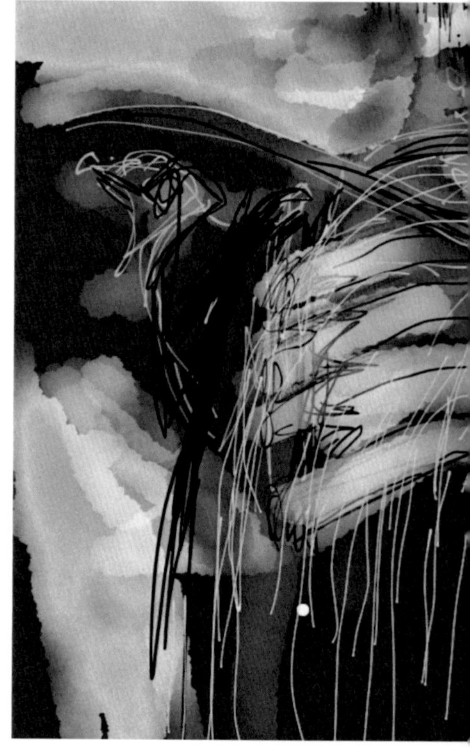

[11] Silvia Plath, "Medusa", in *Ariel* (London: Faber and Faber, 1965), 46.

[12] Karen Barad, "Getting Real: Technoscientific Practices and the Materialization of Reality", *Differences: A Journal of Feminist Cultural Studies* 10, no. 2 (1 July 1998), 119.

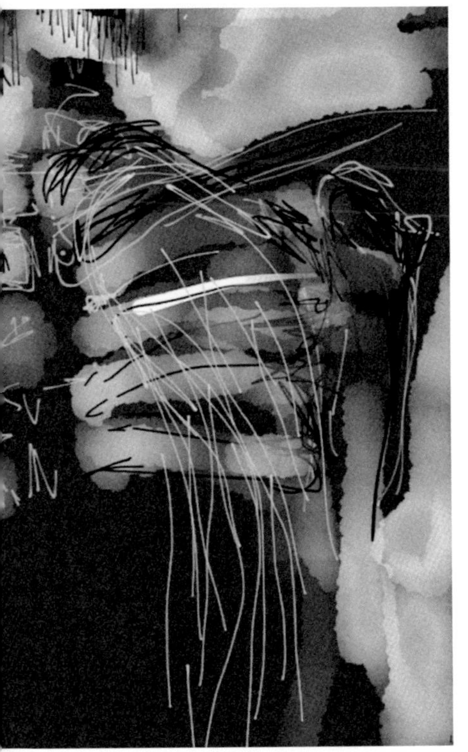

Matters of Staging

[body part] [measurement] [orientations and position, designation of left and right]

[name of the patient] [age, date of birth, gender]

[date, time of the study] [registration number] [social security number of the patient]

[location where the study was performed] [physician's name, initials]

Markers on a scan. Like stamps, or credits of a past event. Witnesses of a past performance to be documented and reconstructed. Credits and recognition to its actors. In the X-ray image, my body enters the scene, transposing experience and emotions, transposing life into data and measurable information. It becomes part of an apparatus of distinction that expands it into a diagnostic, therapeutic and care infrastructure, ready to be processed by social security systems, science, insurance categories, sick notes and prescriptions. As my body becomes evidence, ready to be spectated, it appears simultaneously as witness, document, and agent. Changing registers as it moves across hardly detectable borders drawn by vectors of power and privilege such as class, race, and normalcy.

Projection

These inscriptions, whether writing or drawing, are an offer for under the skin. However, the X-ray is unruly and suffers from being made of both hard corners and fluid boundaries. As a diagnostic tool, the success of reading its milky edges is sometimes tricky. The ache of digital dirt occludes, where digital dirt

> may manifest as baroque accretions or as residual contaminants. Tiny bits of geometrical grit ... Formless lumps may be prolapsed through the photographic surface of representation.[13]

There is also a smear of physical dirt laid over the image, of a kind of violence that corresponds to its digital twin. The dusky residue of mist and dirt, of a quiet violence.
This resistance to being looked at, looked through, also represents a dare: a dare where I attempt a close reading, an audiencing of the X-ray.

[13] Louisa Minkin, "Out of Our Skins", *Journal of Visual Art Practice* 15, no. 2–3 (1 September 2016), 118.

And so, going to its knife edge, I slide the acetate under the hyponychium of my thumb, situated in that available edge between the nail plate and the skin of the fingertip. Settling the flexible material in my hands enhances the sensation of looking in order to peel back layers, to find the borderlands of the X-ray. Holding it away from my body but towards the light it Is both proximal and distal. I am guarding myself against the injuries depicted, however, held up against the daylight I am none the wiser, each dirty corner reveals only a smudge of understanding. This is a light-box-looking where the ghosts of the trauma has a place to return to. The shifting of the bones creak. X-rays of my teeth chatter, the exposure of being outside of the warm mouth leaves them isolated from their nest. X-rays of my arms shatter, and I remember the naming of the scaphoid bone comes from the Greek *skaphos*, which means "a boat", and "boat-like," the most certain knowledges are drifting away from me. The inside-outside sensation of my own body feels confused. The acetate is now warm. I want to taste the X-ray, but I do not hold the X-ray in my mouth. I am frightened away by the stories of women who worked in factories early in the 20th century painting glow-in-the-dark watch faces. They would dip their brushes into radium, then lick the tip of the brushes to create a precise point in order to paint numbers on watch dials; radium poisoning slowly killing them.

I hold these women in my mouth instead.

Exposure

I'm sorry to bother you, but in the event that "X-ray imaging cannot easily visualize soft tissues,"[14] we may need ways to read these bodies that remain. A kind of 'bothering' of technological apparatus, and one which engages in a "politics of cramped spaces."[15] This un-dark X-ray knowledge, is a kind of knowledge that passes through. The X-ray is not quite part of the person who begat it, rather a thin slice of a body in the state of becoming hybrid. It flickers across borders of familiar/strange, of self/other, of inside/outside the body. It is a remembering and forgetting, remembering and forgetting: falling through the gaps of greenstick fractures and grey soft tissue.

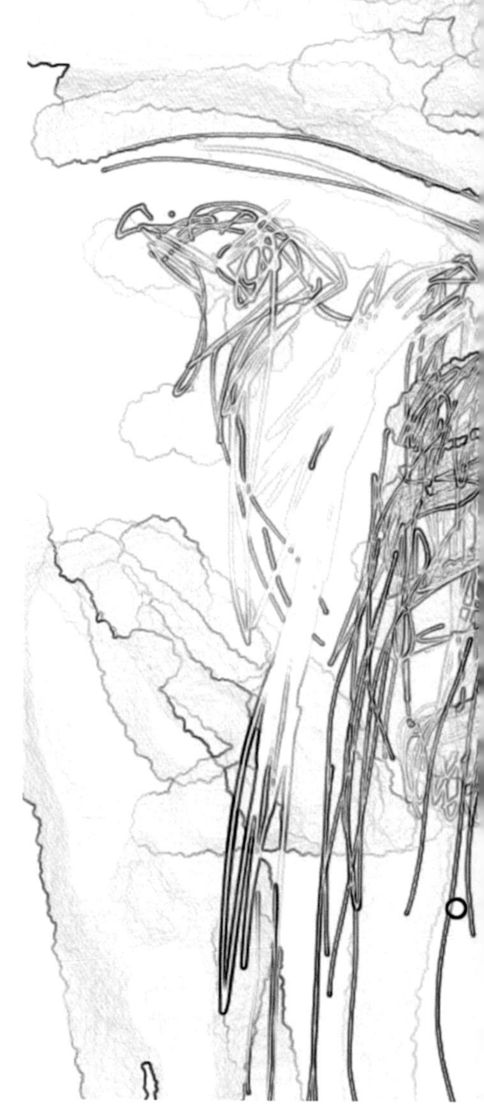

[14] Masami Ando et al., "X-Ray Dark-Field Phase-Contrast Imaging: Origins of the Concept to Practical Implementation and Applications", *Physica Medica*: 79 (November 2020), 189.
[15] William Walters and Barbara Lüthi, "The Politics of Cramped Space: Dilemmas of Action, Containment and Mobility", *International Journal of Politics, Culture, and Society* 29, no. 4 (1 December 2016), 359.

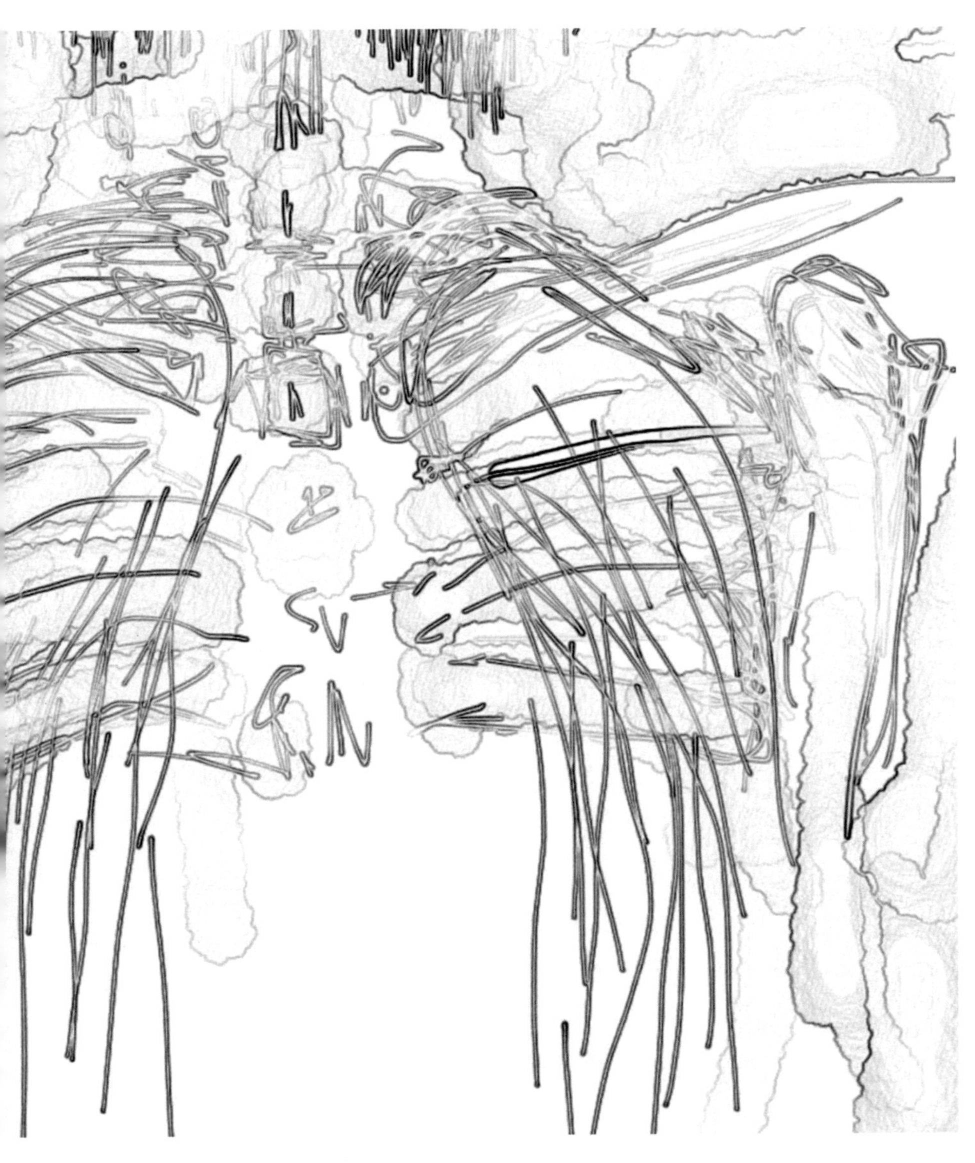

References

Ahmed, Sara. *Queer Phenomenology: Orientations, Objects, Others*. Durham, NC: Duke University Press, 2006.

Ando, Masami, Rajiv Gupta, Akari Iwakoshi, Jong-Ki Kim, Daisuke Shimao, Hiroshi Sugiyama, Naoki Sunaguchi, Tetsuya Yuasa, and Shu Ichihara. "X-Ray Dark-Field Phase-Contrast Imaging: Origins of the Concept to Practical Implementation and Applications". *Physica Medica: PM: An International Journal Devoted to the Applications of Physics to Medicine and Biology: Official Journal of the Italian Association of Biomedical Physics (AIFB)* 79 (November 2020), 188–208. https://doi.org/10.1016/j.ejmp.2020.11.034.

Barad, Karen. "Getting Real: Technoscientific Practices and the Materialization of Reality". *Differences: A Journal of Feminist Cultural Studies* 10, no. 2 (1 July 1998), 87–128. https://doi.org/10.1215/10407391-10-2-87.

Deleuze, Gilles, and Felix Guattari. *A Thousand Plateaus: Capitalism and Schizophrenia*. Translated by Brian Massumi. 2nd edition. Minneapolis: University of Minnesota Press, 1987.

Haraway, Donna J. *When Species Meet*. Illustrated edition. Minneapolis: University of Minnesota Press, 2007.

Juren, Anne. "Lesson on the Skin, Plastering the Body". In Sandra Noeth (Ed.), *What Does It Take to Cross a Border? A Reader*. Berlin: ifa-gallery, 2019.

Lippard, Lucy R. *The Lure of the Local: Senses of Place in a Multicentered Society*. New York, NY: The New Press, 1998.

Lupton, Deborah. *Medicine as Culture: Illness, Disease and the Body*. 3rd ed. London: SAGE, 2012. https://doi.org/10.4135/9781446254530.

Minkin, Louisa. "Out of Our Skins". *Journal of Visual Art Practice* 15, no. 2–3 (1 September 2016), 116–26. https://doi.org/10.1080/14702029.2016.1228820.

Nancy, Jean-Luc. *Being Singular Plural*. Translated by Anne O'Byrne and Robert Richardson. Stanford: Stanford University Press, 2000.

Parkinson, Chrysa. walk + talk 18. Transcription of the performance lecture, Kaaistudio's Brussels, 18 March 2011), available online: http://olga0.oralsite.be/oralsite/pages/Chrysa_Parkinson/.

Plath, Silvia. "Medusa." In *Ariel*. London: Faber and Faber, 1965.

Reid, John A. "The 'visibility' of x-rays". *Oral Surgery, Oral Medicine, Oral Pathology* 34, no. 2 (1 August 1972), 330–34. https://doi.org/10.1016/0030-4220(72)90426-4.

Steinberg, Philip, and Kimberley Peters. "Wet Ontologies, Fluid Spaces: Giving Depth to Volume through Oceanic Thinking". *Environment and Planning D: Society and Space* 33, no. 2 (1 April 2015), 247–64. https://doi.org/10.1068/d14148p.

Walters, William, and Barbara Lüthi. "The Politics of Cramped Space: Dilemmas of Action, Containment and Mobility". *International Journal of Politics, Culture, and Society* 29, no. 4 (1 December 2016), 359–66. https://doi.org/10.1007/s10767-016-9237-3.

Sandra Noeth **and** **Joanne 'Bob' Whalley**

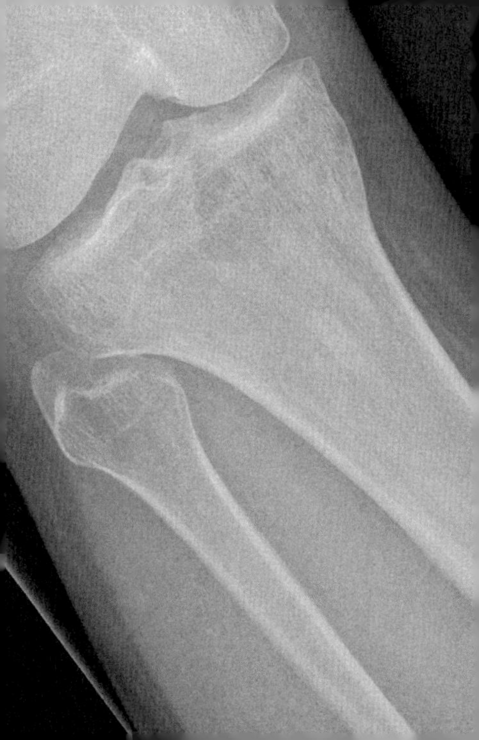

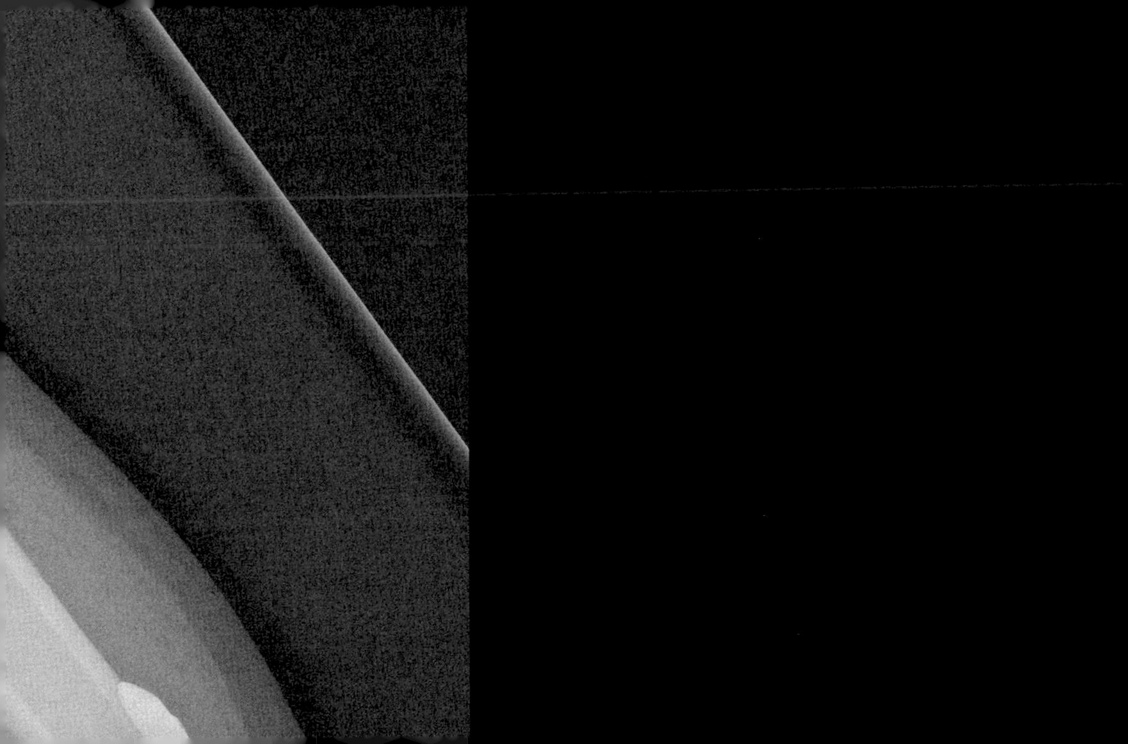

Shahram Khosravi

Listening to X-ray Images. A Family Album

“This writing is personal because this history has engendered me, because ‘the knowledge of the other marks me,’ because of the pain experienced in my encounter with the scraps of the archive, and because of the kinds of stories I have fashioned to bridge the past and the present and to dramatize the production of nothing—empty rooms, and silence, and lives reduced to waste.” — Saidiya Hartman[1]

My interest in X-ray photos has been shaped by personal tragedy. After a stroke in the spring of 2020, my father lost the ability to speak. We did not know each other very well. When I was a teenager, he was in prison or in exile, and in 1987, when I was twenty years old, I escaped from Iran. For most of my time in exile, I could not return and he was subject to a travel ban. Distance led to silence. Apart from formal and very short chats on *Norouz*, the Iranian New Year, we never spoke. My father passed away two years after the stroke. In those two years, he could not move and had severe expressive aphasia. After the stroke, I went to Iran to find him struggling to find words. Sitting beside his bed, taking his hand in my hands for the first time since childhood, I saw how he suffered from not being able to communicate. I found many X-ray photos of his brain and chest in his house. I tried to find words inside him, words that had been buried in his chest and stories that he never had a chance to tell me. There are several Persian expressions that say the chest is a space of hidden and unspoken words; such as *sine malamal dard ast* (the chest is brimful of pain) or the act of *dard-e del* (telling someone about the grief in your heart).

[1] Saidiya Hartman, “Venus in Two Acts,” *Small Axe: A Caribbean Journal of Criticism* 12, no. 2 (1 June 2008), 4. https://doi.org/10.1215/-12-2-1.

When I look at the X-ray photos of my father's chest, I do not look; I listen and am always touched by their vibration. This is an attempt to connect with him across time. Tina Campt uses the term "haptic temporalities" to show us the multi-layered and complex ways an image signifies through vision, touch, and sound. X-ray photos became the only medium for me to reach him and to imagine what he would have said, if only he could speak.[2] I approach the X-ray photos of my father as an archive of violence, political oppression, and bordering practices. It has become the beginning of my search for other X-ray photos, other bodies, other stories.

[2] Tina Campt, *Listening to Images* (Durham: Duke University Press, 2017).

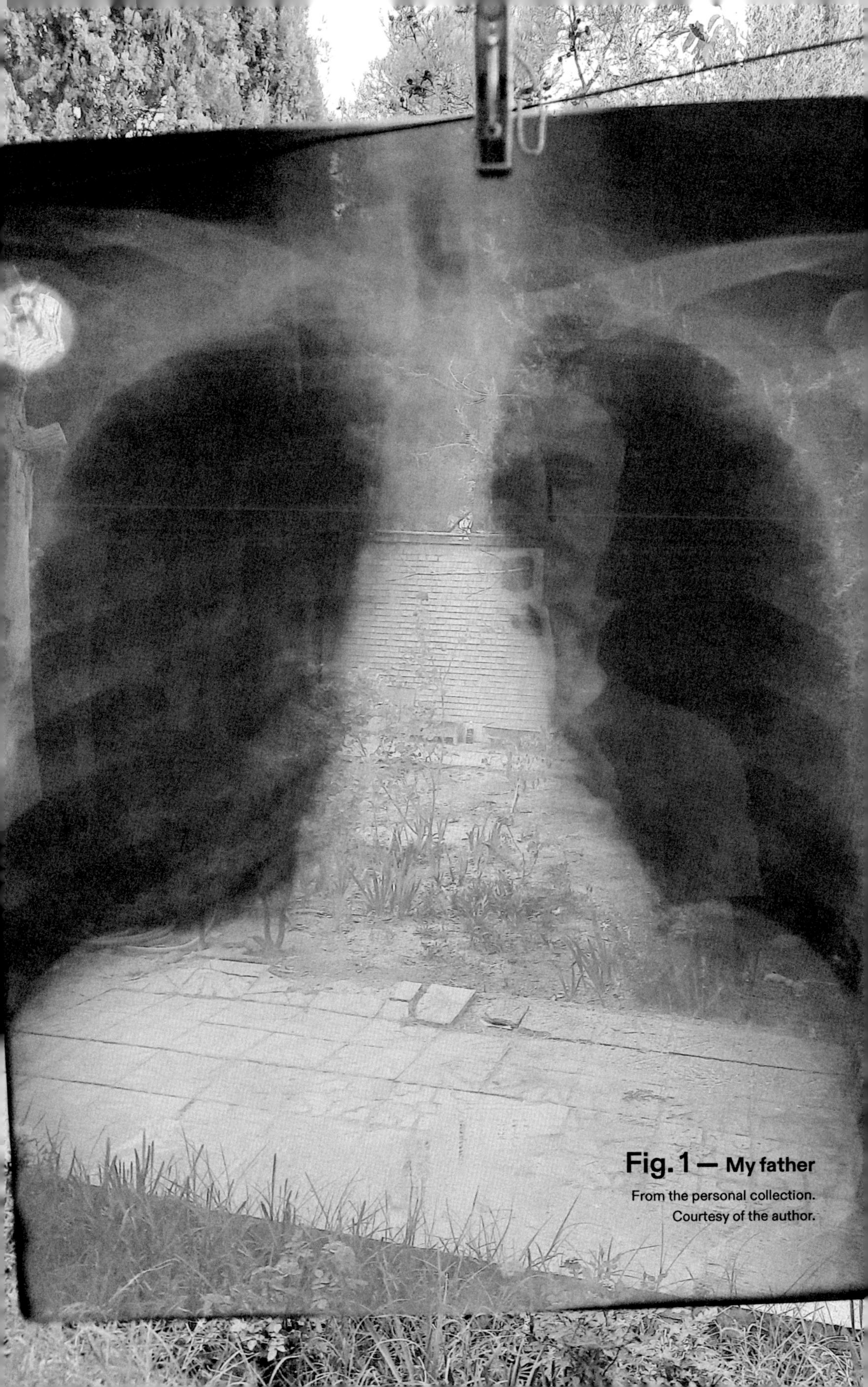

Fig. 1 — My father

From the personal collection.
Courtesy of the author.

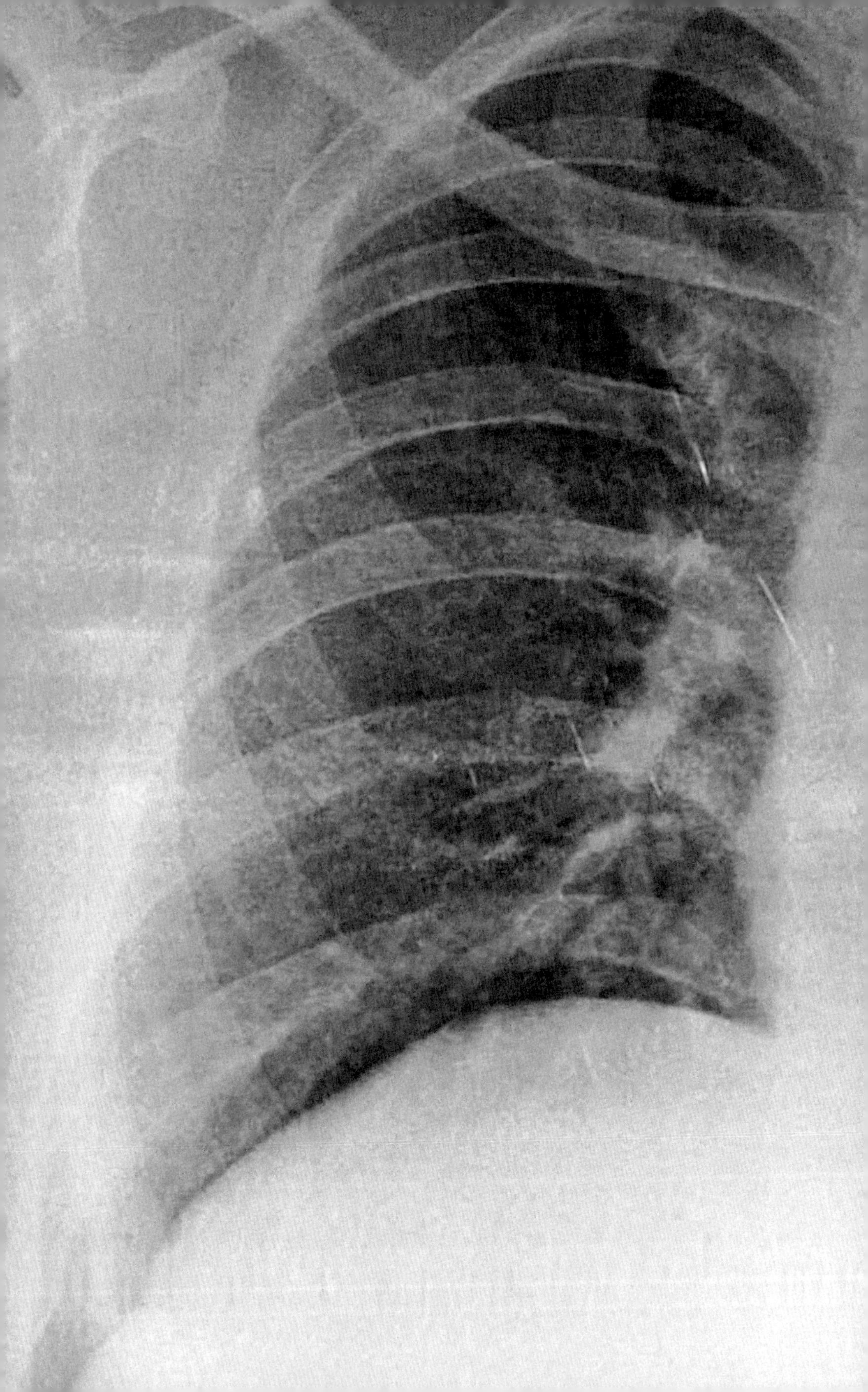

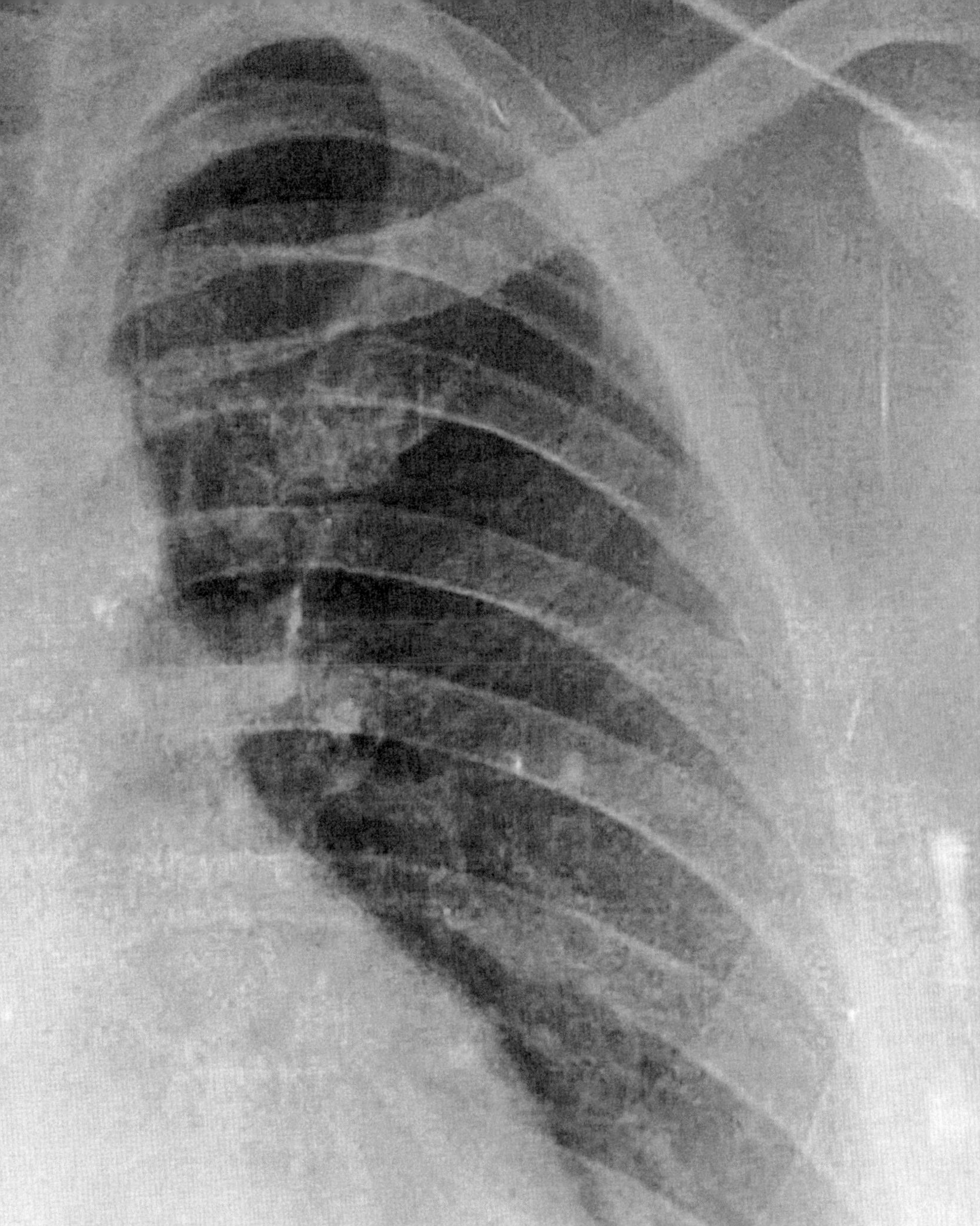

Fig. 2 — This is Monish, a friend from New Delhi. Some years ago, he applied for a visa for Qatar. As part of the visa procedure, he had to do a tuberculosis test. The doctors in Qatar suspected signs of TB so his visa application was rejected. Worried for his health, he did another X-ray test in a local clinic in New Delhi. Doctors saw no traces of any health problems. However, after several years and with no chest problem, he still is anxious about what doctors in Qatar had seen in his chest X-ray photos. From the personal collection. Courtesy of Monish.

Fig. 3 — This is Yousef. The first time I met him, he was in a detention centre in Stockholm before being deported to Afghanistan after his asylum application was rejected. It was the winter of 2012. After a year or so in Kabul, he crossed the border and became an undocumented migrant in Iran. His cousin found a job for him as street sweeper in a private company. Contracted by a municipality in south Tehran, the company sent him and other migrant workers to clean the streets after midnight. Without any high-visibility jackets or reflective markings, he was totally invisible in the dark. One night, about two hours past midnight, he was knocked over by a car. Severely injured, he was hospitalised for two weeks. Hardly able to walk, the police picked him up and he was deported to Afghanistan. Several years later in May 2020, he tried to cross the Iranian border again. Unfortunately, he was shot by border guards in the same foot which was crushed in the car accident. With a foot smashed twice, he can't walk more than a few steps. He has been stuck in a small town in western Afghanistan, washing dishes in restaurants. From the personal collection. Courtesy of Yousef.

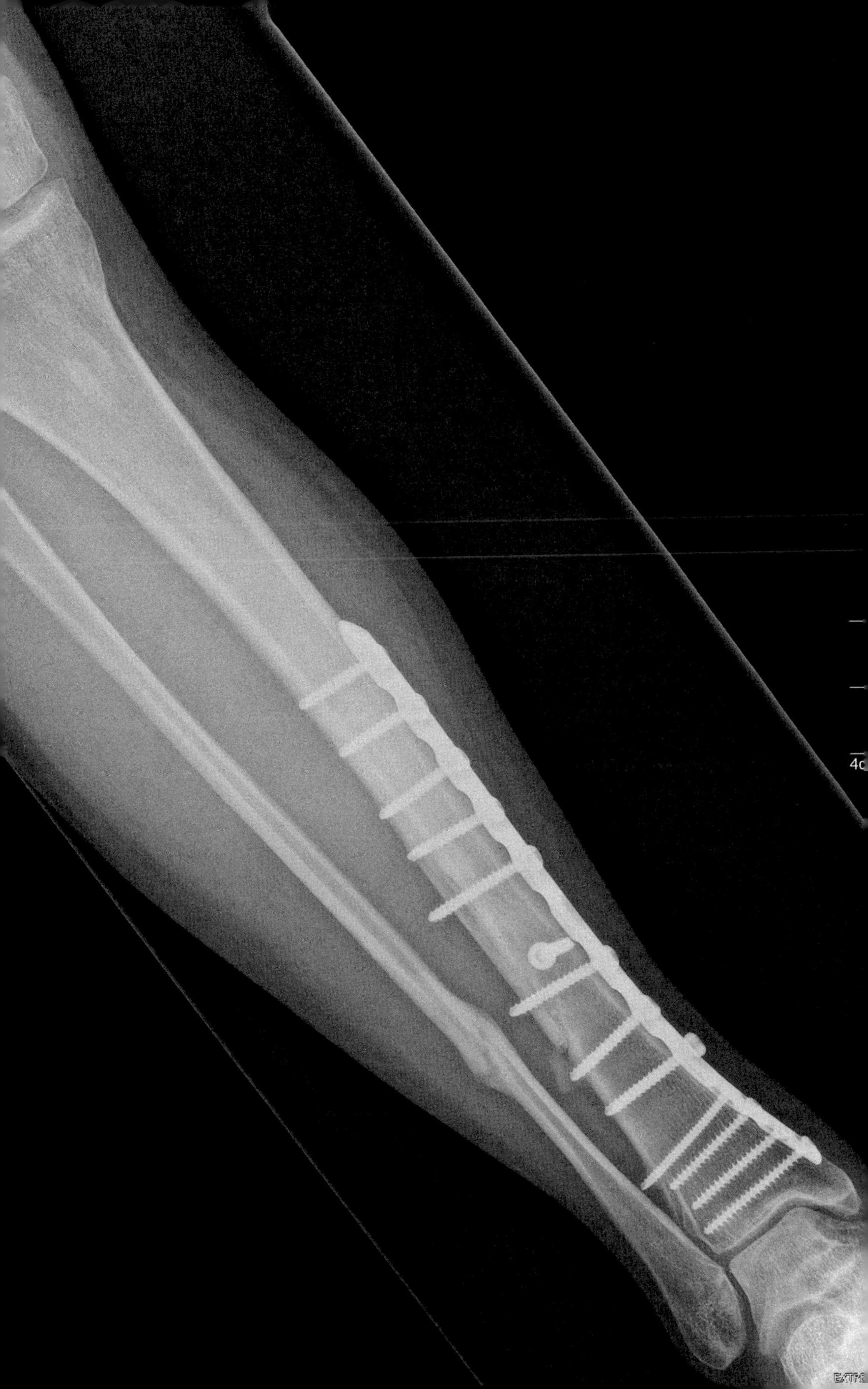

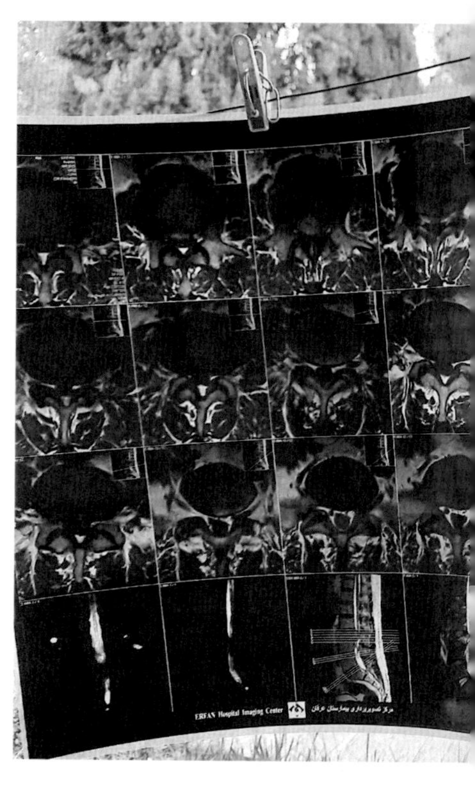
ERFAN Hospital Imaging Center
مرکز تصویربرداری بیمارستان عرفان

Fig. 4 — **This is Abbas, a childhood friend from my village in the Bakhtiari region in south west Iran. He was forced to join the army during the Iran-Iraq War (1980–1988). In early 1986 at the front, he was injured by chemical weapons used by the Iraqi army. The repeated screenings of his chest show that his lungs are seriously damaged. His skin is burnt and he suffers from depression. He never received proper treatment either during or after the war. Today, he moves from one manual labour job to another. His damaged lungs force him to sit down to take a breath after a short walk. His burnt skin makes working in the sun unbearable. The state classified him as the lowest rate of war-disabled veteran, which is the 10 percent rate. For 10 percent, he received an insignificant compensation. The X-ray photos are the evidence that makes him 10 percent: 10 percent included; 10 percent citizen.** From the personal collection. Courtesy of Abbas.

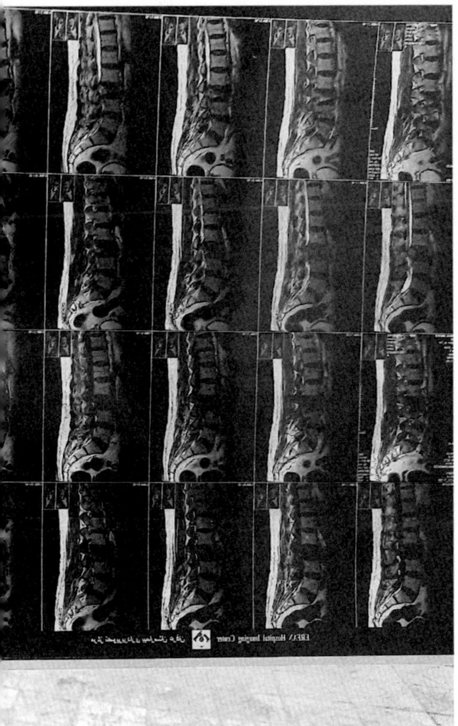

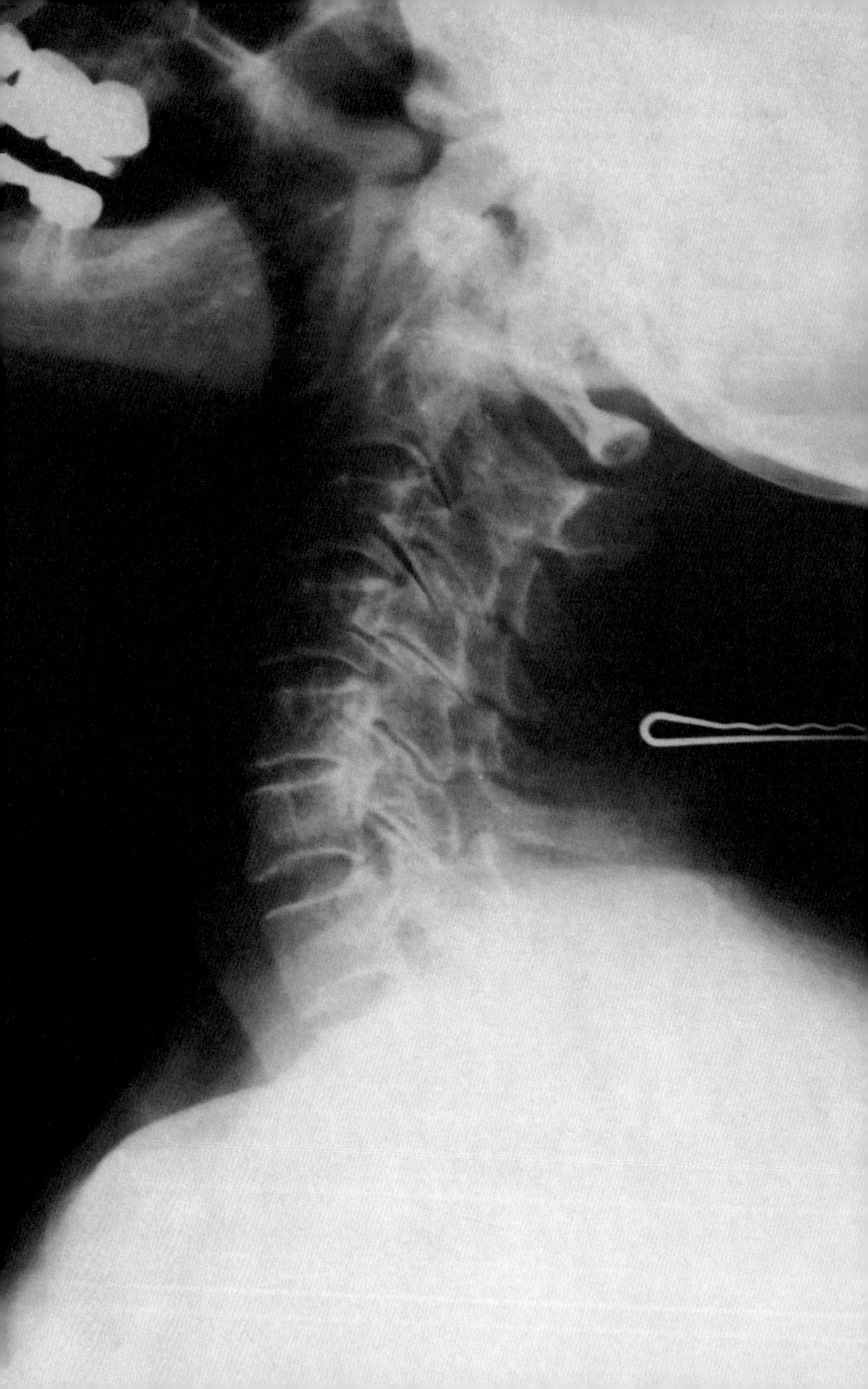

Fig. 5 — **My mother. She worked for 61 years, since she was 18 years old. Every single day for 61 years. Her work was informal: 61 years of unregistered and undocumented labour. A long working life rendered completely invisible. After 61 years without insurance and pension the only thing she was left with was an excruciatingly painful and worn-out body. In the closet next to her bed was a load of X-ray photos from all over her body. I chose this one because of the hairpin she had forgotten to remove before being X-rayed. This is how I remember her. The hairpin used to hold her long black hair in place, making her ready to work with her hands. The hairpin also adds a personal feature to the X-ray photo, which is otherwise cleared from all traces of her as a person.**

From the personal collection. Courtesy of Shahram Khosravi.

Fig. 6 — This is me. In October 1991, I was shot by the Laser Man (Lasermannen, in Swedish), a racist terrorist who used guns equipped with laser sights to target migrants in Stockholm. He shot 11 people between August 1991 and February 1992. That cold and dark Scandinavian night, a bullet bored through my cheek on the right side of my face, hit my teeth and split inside my mouth. Since removing the remaining bullet fragments would have severely damaged my cheek, the doctors decided to leave them in my cheekbone. Bullet fragments are still lodged deep in my bones—souvenirs of the border gaze.

From the personal collection. Courtesy of Shahram Khosravi.

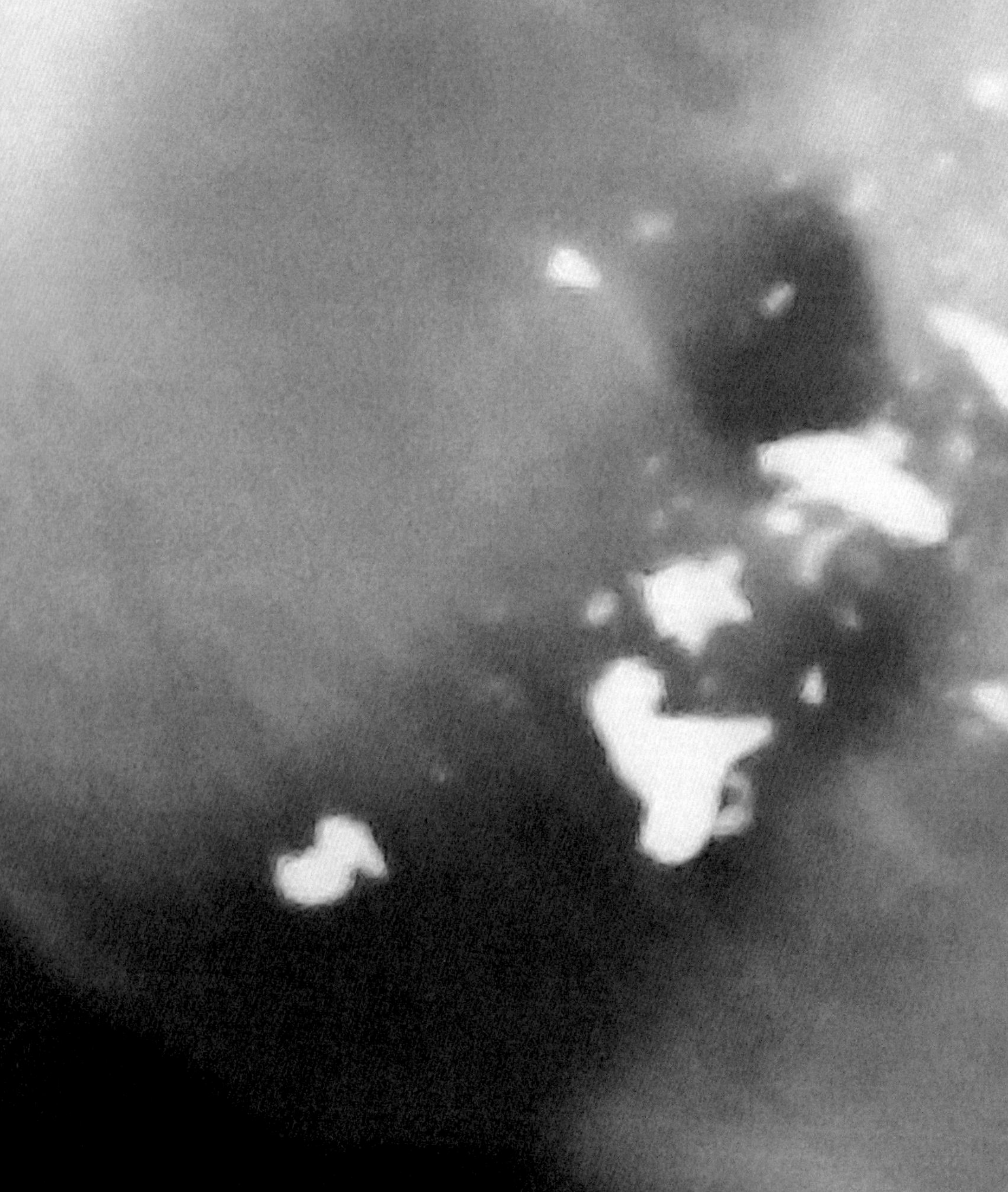

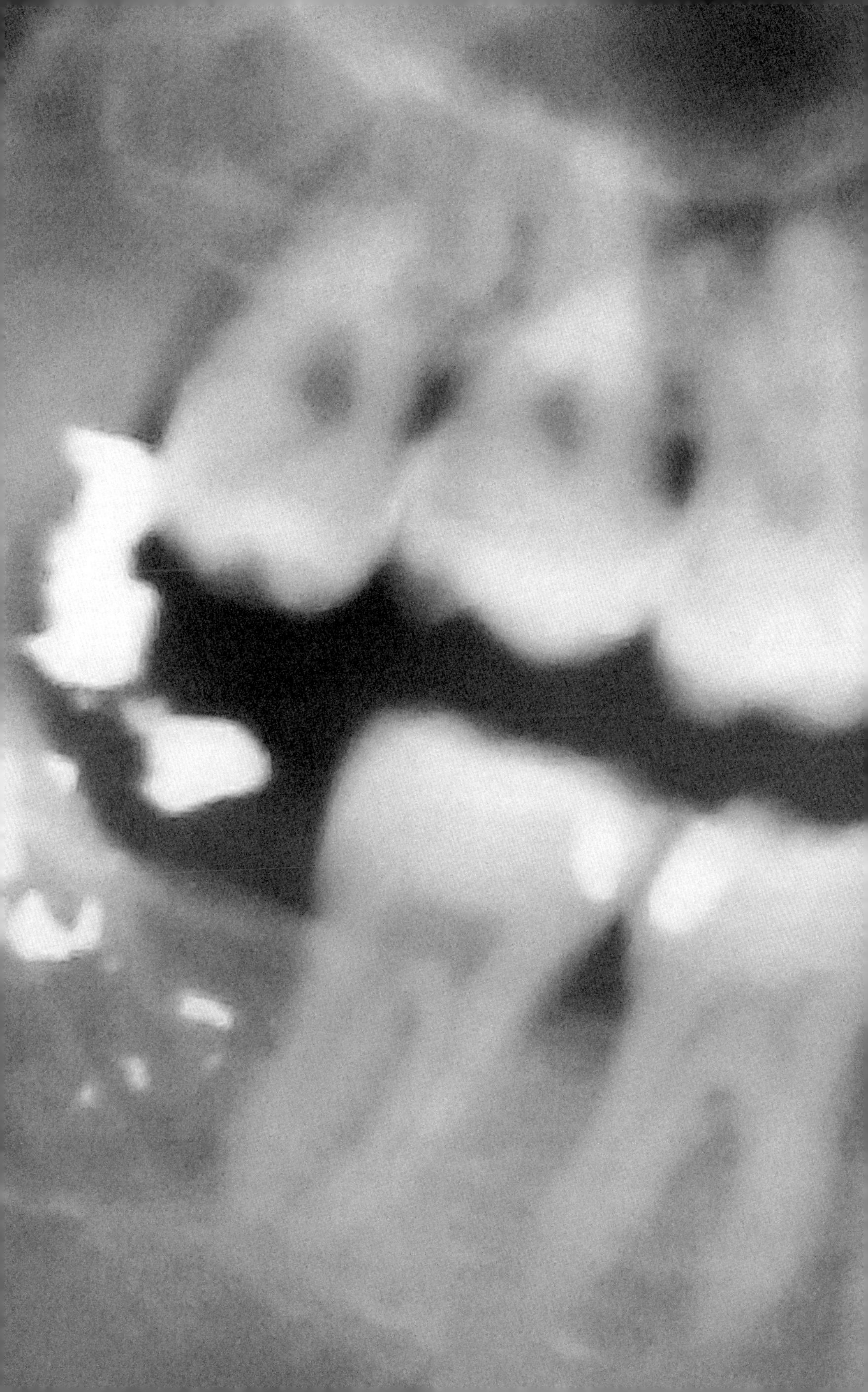

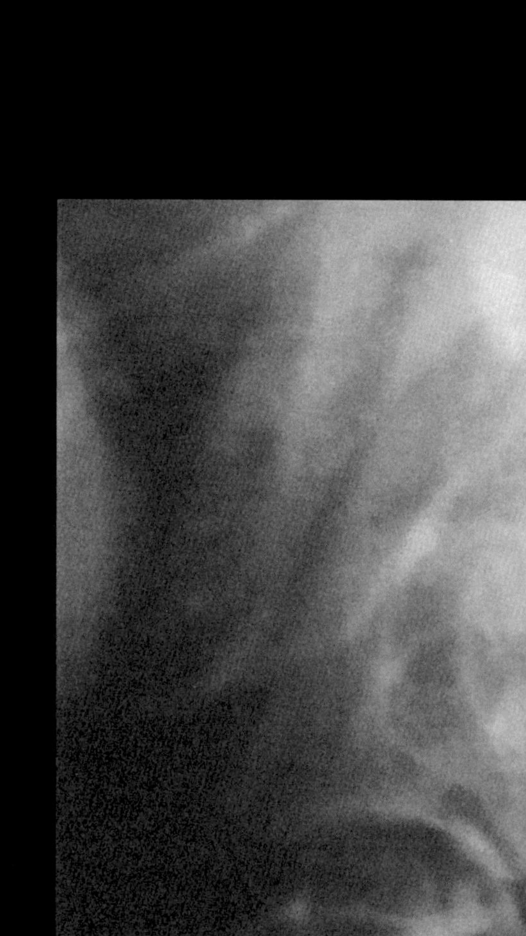

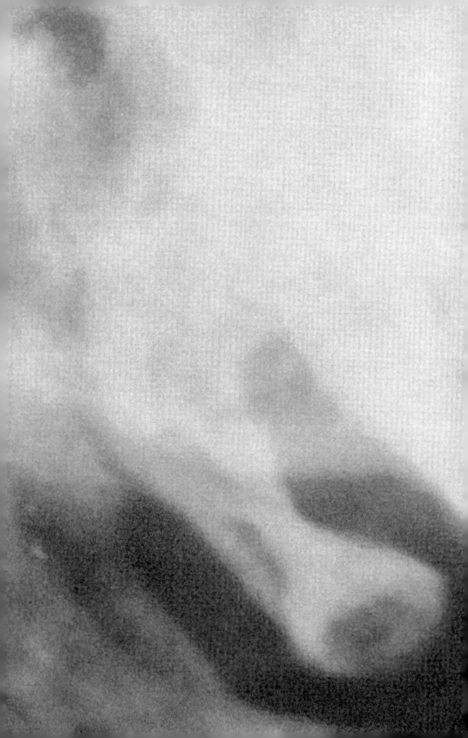

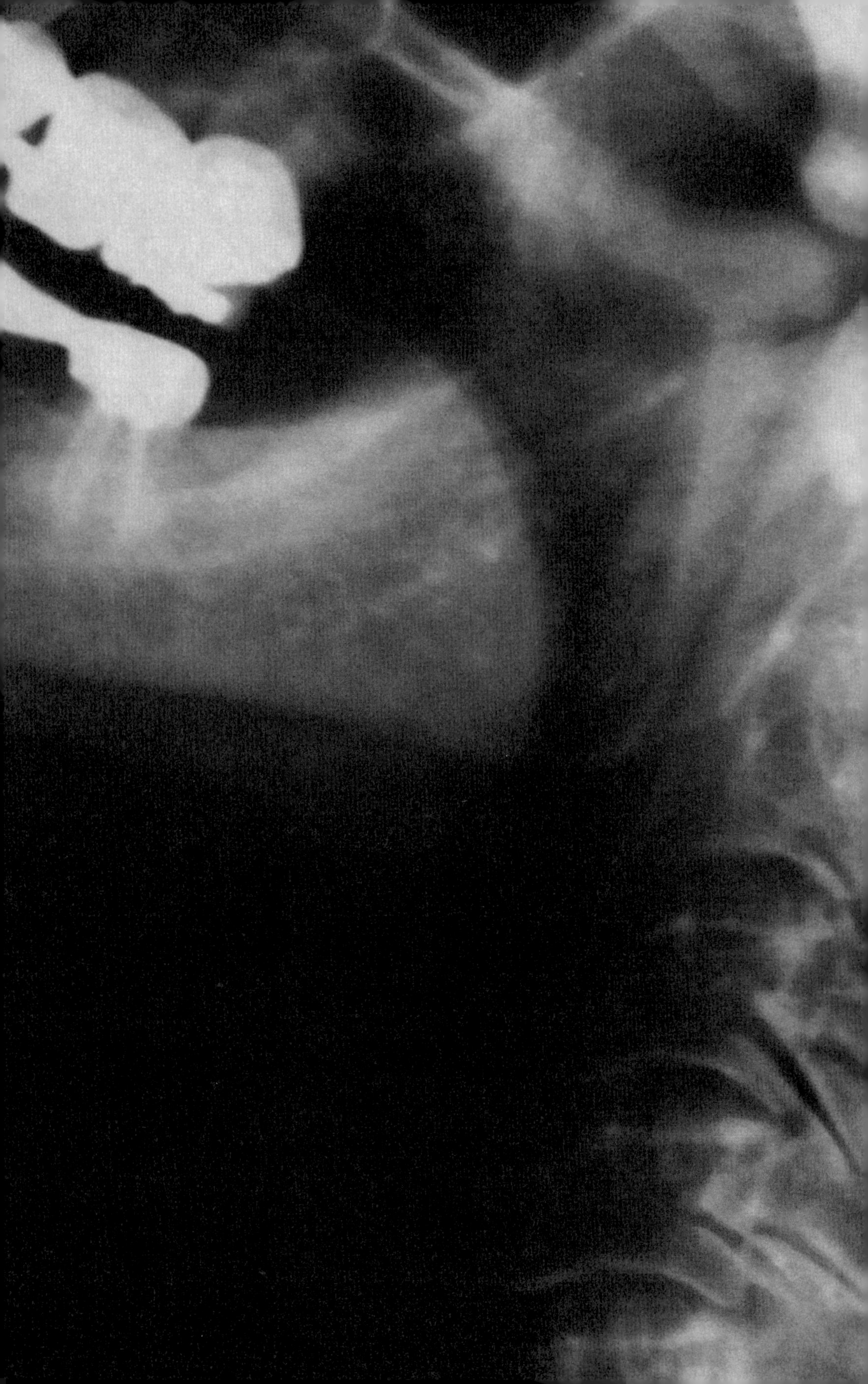

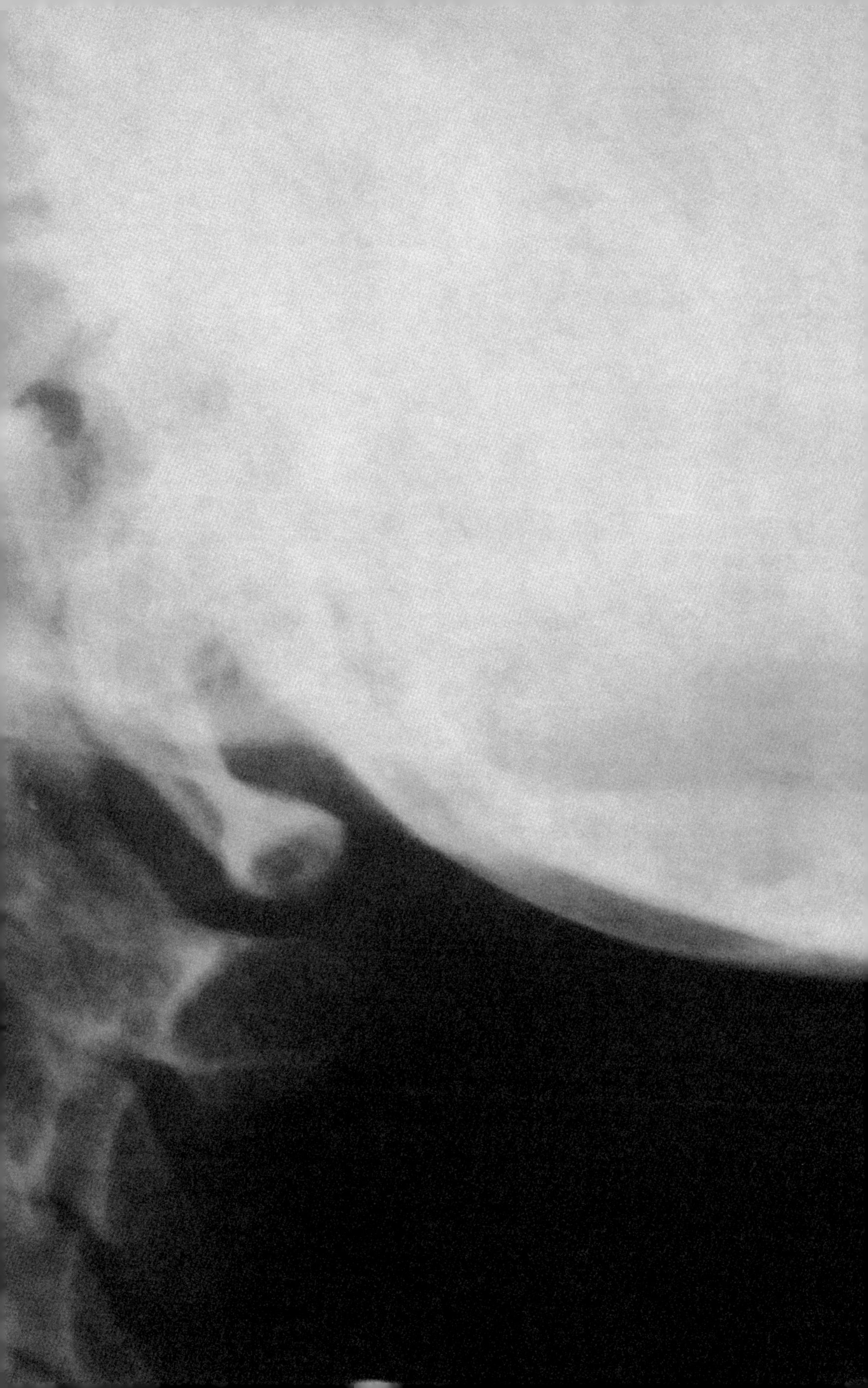

Contributors

Javad Abbasi Tavallali is an Iranian journalist and human rights activist.

Paolo Favero is an anthropologist with an interest for the meaning of images in human life. Currently Professor of Visual Anthropology and Cultures at the University of Antwerp he conducts research on the nexus between visual culture and existential matters. His present work explores the nexus between matters of dying and mourning and aesthetics across cultures. An active image-maker Paolo also employs emerging digital technologies (among them smartphone apps, immersive cameras and drones) in the terrain of ethnographic methods and art. Paolo is the author of *Image-Making-India* (Routledge, 2020) and *The Present Image* (Palgrave Macmillan, 2018).

Ella Hillström is a Ph.D. student in social anthropology at Stockholm University. She is an interdisciplinary writer with a dance-thinking practice together with Yari Stilo. She taught at Parsons School of Design after completing her M.A. in Anthropology from the New School for Social Research.

Reza Hussaini is a researcher from Afghanistan and currently a PhD candidate at City, University of London.

Shahram Khosravi is professor of Anthropology at Stockholm University. His research interests include the anthropology of Iran, forced displacement, border studies and temporality. Khosravi is the author of several books including *Young and Defiant in Tehran* (2008); *The Illegal Traveler: an auto-ethnography of borders* (2010); *Precarious Lives: Waiting and Hope in Iran* (2017); *After Deportation: Ethnographic Perspectives* (2017, editor); *Waiting. A project in Conversation* (2021, editor), and *Seeing Like a Smuggler* (2022, co-editor). He has been an active writer in the international press. He is a co-founder of Critical Border Studies, a network for scholars, artists and activists to interact.

Behzad Khosravi Noori is an artist, writer, educator, playground maker, and necromancer. His practice-based research includes films, installations, and archival studies. His works investigate histories from the Global South, labour and the means of production and histories of political relationships that have existed as a counter narration to the east-west dichotomy during the Cold War and beyond. By bringing multiple subjects into his study, Behzad explores possible correspondences seen through the lenses of contemporary art practice, proletarianism, subalternity, and the technology of image production. He analyses recent history to revisit memories beyond borders, exploring the entanglements and non/aligned memories.

Jonathan Krämer is a doctoral candidate at the Department for Social Anthropology at Stockholm University. His research concerns the role of medical examinations in transnational labour migration in Southeast Asia, having done fieldwork Indonesia and Malaysia. He has previously been a Visiting Fellow at the Max Planck Institute for Social Anthropology in Halle, Germany.

Sandra Noeth is a professor at HZT Berlin and internationally active as a curator. She specializes in ethical and political perspectives of body practice and theory. Her selected research projects include: *Violence of Inscriptions*, a program series on bodies under structural violence (with A. Zaides, 2016-18, HAU,Hebbel am Ufer); *What Does it Take to Cross a Border?* (2019, ifa-gallery Berlin); *Bodies, un-protected* on the relation between bodies, art and protection (2020-22, with Künstler*innenhaus Mousonturm); *Hållning*, a practice-led platform for collective learning (2021, with MDT Stockholm). Sandra acted as Head of Dramaturgy and Research at the Tanzquartier Wien between 2009-14, where she co-edited the periodical *SCORES*. Further publications include: *Breathe. Critical Research into the Inequalities of Life* (2023, transcript, co-editor with J. Janša); *Violence: Embodiment* (2022, editor, PARSE journal, online); *Bodies of Evidence: Ethics, Aesthetics, and Politics of Movement* (2018, Passagen, co-editor with G. Ertem) and *Resilient Bodies, Residual Effects: Artistic Articulations of Borders and Collectivity from Lebanon and Palestine* (2019, transcript).

Ayodamola Tanimowo Okunseinde is a Nigerian-American artist, designer, and time traveller living and working in New York. Okunseinde holds a BA in Visual Arts from Rutgers University, an M.F.A. in Design and Technology, and an M.A. in Anthropology from The New School. He is currently a Ph.D. student in Anthropology at The New School for Social Research and serves as an Assistant Professor of Interaction and Media Design at Parsons School of Design.

Yousif M. Qasmiyeh was born and educated in the Baddawi refugee camp in Lebanon. He is a scholar and poet whose Dr. phil. research at the University of Oxford explores containment, the archive, and time in refugee writing. He is writer-in-residence for the AHRC funded Refugee Hosts project and the joint lead of the Baddawi Camp Lab as part of the Imagining Futures GCRF-Network+ project. His essays, poetry and translations have appeared in *Modern Poetry in Translation, Critical Quarterly, GeoHumanities, Cambridge Literary Review, PN Review, Stand, New England Review, Poetry London* and the *Journal of Refugee Studies*. His collection, *Writing the Camp* (2021, Broken Sleep Books), was a Poetry Book Society recommendation and was selected as one of the Best Poetry Books of 2021 by *The Telegraph* and the *Irish Times*; was highly commended by the Forward Prizes; and was shortlisted for the 2022 Royal Society of Literature Ondaatje Prize. His latest book is *Eating the Archive* (2023, Broken Sleep Books).

Partha Sengupta is an Indian activist and award-winning documentary photographer. Born in Calcutta to a refugee family that migrated from East Pakistan, now Bangladesh. In 2012, he left his corporate job and embraced photography full time, joining the staff of a local newspaper, and working there for three years. During the Counter Foto mentorship program in Bangladesh, he began exploring work about the Bengal Partition. This would result in *The Bloodiest Border* and was a finalist for *Burn* Magazine's Emerging Photographer Award. In 2018, the Serendipity Arts Foundation Fellowship to develop *Sons of Soil*, about Bengali refugees who migrated, like his family, to India in 1971 and was exhibited at Goa, Kozhikode during the Sub-Altern Festival, at the Indian Parliamentary Gallery. In 2022, he was awarded a full scholarship to study Documentary Photography and Visual Journalism at The International Centre of Photography in New York. In 2023, his images of border violence were exhibited at the Angkor Photo Festival in Cambodia and the finalist in the Chennai Photo Biennale. He is currently working independently and based in Calcutta.

Yara Sharif is a practising architect and academic with an interest in design as a means to rethink contested landscape, with a new take on architectural practice. Her research, design and pedagogical practice aim to disrupt the colonial power with its hegemony and hierarchy of knowledge production, through alternative spatial narratives. She sheds a light on "forgotten" communities, while also interrogating the relationship between politics and architecture. Her book *Architecture of Resistance* as well as the collaborative work with PART on re-imagining spatial possibilities in Palestine, have won the RIBA President's Award for Research in 2013 and 2016.

Françoise Vergès (Reunion Island) is currently Senior Research Fellow, Sarah Parker Centre for the Study of Race, UCL, an antiracist feminist activist, independent curator and writer. Recent publications include *A Feminist Theory of Violence* (2022), *A Decolonial Feminism* (2021) and *Programme de désordre absolu. Décoloniser le musée* (2023, forthcoming in English).

Joanne 'Bob' Whalley is a Reader in Performance, and Director of Doctoral Training and Development, at University of the Arts London, UK. Her book *Between Us: Audiences, Affect and the In-Between* (Palgrave Macmillan, 2016), published with her research partner Lee Miler, celebrates spaces which cause an affecting, and bodies affected. In 2015 she completed a B.Sc. in Acupuncture, and she specialises with/through intersectional narratives. Her Ph.D. students explore grief narratives, empathy and affective exchange, concepts of with-ness and witness. Whalley and Miller completed the first joint practice-as-research Ph.D. to be undertaken within a UK arts discipline in 2004, and they make performance, installation, performance text and objects for international audiences.

The team of the publication series *Corporeal Matters*

Daniel Belasco Rogers

was born in London in 1966. He has worked as a designer, director, composer and performer for experimental theatre and performance since 1989. After moving to Berlin, he established the artist duo *plan b* in 2002, together with Sophia New. As well as making durational performances, installations, social interventions and new media projects, Sophia and Daniel have a practice of recording every journey they make with a GPS. Daniel was a member of the Junge Akademie at the Akademie der Künste in 2006 and has taken part in artist residencies in Germany, Austria, the UK and Norway. Daniel has held practical artistic workshops in Tokyo, Sao Paulo, Banff (Canada), Beijing and many other European cities and taught as a guest lecturer in higher educational institutions across Europe. As a performer and collaborator, he has worked with She She Pop, Club Real, Gob Squad, Forced Entertainment, Sabine Zahn and Juan Domínguez/ Arantxa Martínez. He was a guest professor at the University of the Arts Berlin (2020–2023). https://planbperformance.net

Janez Janša

is a professor at HZT Berlin and a contemporary artist who focuses on the relationship between art and the social and political context in his performance, conceptual and interdisciplinary art works. His particular areas of research are the performativity of name, the relationship between art and war, as well as time and temporalities in art and life. He was the director of *Maska* (1998–2021), an institute for publishing, artistic production and education based in Ljubljana, Slovenia, and founder and editor of two book series and several readers on contemporary dance and theatre. He was editor-in-chief of *Maska*, performing arts journal (1999–2006). He is the author of a book on Jan Fabre's early work: (*La discipline du chaos, le chaos de la discipline*, 1994). Janez is a co-founder and the first president of the association of freelance artists *Asociacija* in Slovenia and a member of the editorial boards of the journals *Performance Research* and *Maska*. In 2007, together with two other artists, he changed his previous name into the name of the conservative three-time prime minister of Slovenia. Together with Janez Janša and Janez Janša, he is the owner of the Janez Janša® registered trade mark.

Ana Lessing Menjibar

is a German-Spanish performer, choreographer, multidisciplinary artist and art director, born and based in Berlin. In her interdisciplinary practice, she interweaves body, sound worlds and language in multimedia installations in which she experiments with the transformative potential of flamenco in the context of contemporary dance and performance. She graduated from the performance art Master's program, Solo/Dance/Authorship at HZT Berlin. Previously she studied Visual Communication at the University of the Arts Berlin and has worked as an art director and publisher in the field of culture and art for many years. Amongst many other places, Ana Lessing Menjibar has performed, directed or exhibited at Uferstudios Berlin, sophiensaele, Komische Oper Berlin, tanzhaus nrw, and the Kammermusiksaal der Berliner Philharmonie, at Villa Romana (Italy), PHotoEspaña or at the Centre Pompidou Málaga (Spain). https://analessingmenjibar.com

Sandra Noeth
See contributors, page 185.

Sandra Umathum
is a theatre and performance scholar as well as a dramaturge. From 2019 to 2022, she was a Professor for (Applied) Theory Dance, Choreography, Performance at HZT Berlin. Before that, she held a professorship for Theatre Studies and Dramaturgy at the Ernst Busch Academy of Dramatic Arts in Berlin (2013–2018) and a guest professorship for Dramaturgy at the University of Music and Theatre Felix Mendelssohn Bartholdy in Leipzig (2010–2012). She is the author of *Kunst als Aufführungserfahrung*, a book on intersubjective experiences in the visual arts (transcript, 2011), and has co-edited, among other publications, *Disabled Theater* (diaphanes, 2015) and *Postdramaturgien* (Neofelis, 2020). Her research focuses on the theory and practice of contemporary theatre and performance on illness, disability and non-normative bodies in performance, on performance and/as documentation, on shooting (with guns and with cameras), and on new forms of dramaturgy.

Acknowledgements

Working on this book has been difficult. I have been dealing with an archive of violence. Many people, however, helped me bear the unbearable. I thank you all and I am deeply indebted to you for sharing your work in this book.

I thank Sandra Noeth, Sandra Umathum and Janez Janša, the series editors of *Corporeal Matters.* Sandra Noeth believed in this project and without her support and encouragement I am not sure if I would have been able to produce this book. I also thank Anna Wagner and Katja Armknecht at Künstler*innenhaus Mousonturm, Frankfurt. I thank Omar Berrada for hosting my first talk on X-ray photos at The Cooper Union, New York. Thank you Annika Lindberg, Amin Parsa, and Mahmoud Keshavarz for your insightful comments. Thank you Ana Lessing Menjibar for design and layout of the book. Thank you Daniel Belasco Rogers and Matthew Ashton for proofreading.

This book is based on research results that emerged within the framework of *Bodies, un-protected*, a programme of Künstler*innenhaus Mousonturm, funded by the German Federal Cultural Foundation, the Federal Agency for Civic Education (Bundeszentrale für politische Bildung), the Goethe-Institut, the Freunde & Förderer des Mousonturm e.V., the Rudolf Augstein Stiftung and the Hessische Theaterakademie. Supported by the Inter-University Centre for Dance Berlin (HZT).

Imprint

Corporeal Matters

Series Editors
Janez Janša, Sandra Noeth & Sandra Umathum

Book no. 2
THE GAZE OF THE X-RAY
An Archive of Violence

Editor
Shahram Khosravi

Authors
Paolo Favero, Ella Hillström, Reza Hussaini, Shahram Khosravi, Behzad Khosravi Noori, Jonathan Krämer, Sandra Noeth, Ayodamola Tanimowo Okunseinde, Yousif M. Qasmiyeh, Partha Sengupta, Yara Sharif, Javad Abbasi Tavallali, Françoise Vergès, Joanne 'Bob' Whalley

Design Ana Lessing Menjibar
Design Support Peter Löffelholz
Copy Editing Daniel Belasco Rogers
Printed by Majuskel Medienproduktion GmbH, Wetzlar
Print-ISBN 978-3-8376-7048-6
PDF-ISBN 978-3-8394-7048-0
DOI https://doi.org/10.14361/9783839470480
ISSN of series 2941-1688
eISSN of series 2941-1696

Bibliographic information published by the Deutsche Nationalbibliothek
The Deutsche Nationalbibliothek lists this publication in the Deutsche Nationalbibliografie; detailed bibliographic data are available in the Internet at https://dnb.dnb.de/

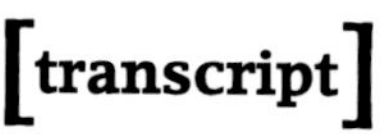

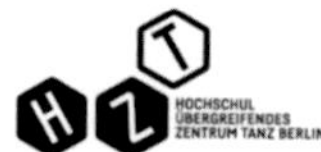

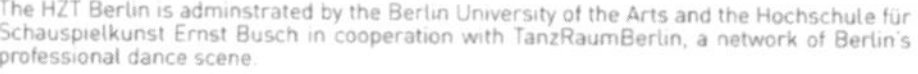

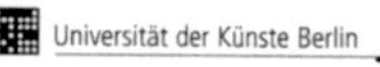

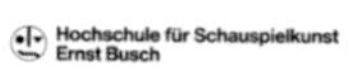

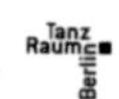